KEY CONCEPT ACTIVITY
LAB WORKBOOK

ALGEBRA FOUNDATIONS:
PREALGEBRA,
INTRODUCTORY ALGEBRA
& INTERMEDIATE ALGEBRA
SECOND EDITION

Elayn Martin-Gay

University of New Orleans

Pearson

ScoutAutomatedPrintCode

Pearson

ISBN-13: 978-0-13-588334-1
ISBN-10: 0-13-588334-2

Table of Contents

Answer Key

Answers are available. Talk to your instructor.

1.   Extension Exercise—Hidden Costs of Planning a Party

Your company is planning an employee holiday party. In addition to the direct costs, such as food and door prizes covered by the budget, the company president wants to know hidden labor costs, or how much the party will cost the company in terms of planning time. You need to help the party committee calculate the hidden labor costs for planning the party. The four employees on the party committee are Karen, Tom, Juan, and Debbie.

a. After subtracting vacations and holidays for the year, Karen, Tom, Juan, and Debbie, each work 48 weeks a year, 40 hours a week. Determine how many hours each employee works in a year.

Hours worked per year per employee: _____

b. The annual salaries for Karen, Tom, Juan, and Debbie are shown in the table below. For each person, determine the hourly cost (in dollars) to the company, or how much one hour of work costs the company. Show your answers in the table.

Hint: hourly cost = yearly pay ÷ number of hours worked.

Employee	Yearly Pay ($)	Calculations	Hourly Cost to Company ($)
Karen	$28,800		
Tom	$30,720		
Juan	$32,640		
Debbie	$36,480		

c. Karen, Tom, Juan, and Debbie met five different times for two hours each to plan the party. Each meeting was held around lunchtime, so that one hour of the meeting was during the lunch hour (not on company time) and the other hour was on company time. How many hours of company time did each committee member spend planning the party? Show your work.

Hours of company time each committee member spent planning: _____

d. In the table below, copy the hourly cost for each employee (from part *b*). For each of the committee members, record the number of planning hours on company time (from part *c*), and determine how much the planning meetings cost the company in lost wages.

Employee	Hourly Cost ($) (copy from part *b*)	Hours on Company Time (copy from part *c*)	Calculations	Cost to Company ($)
Karen				
Tom				
Juan				
Debbie				

e. Explain why the direct costs of a party, such as food and drinks, decorations, and prizes do not necessarily cover all of the costs when a company plans a party for its employees. Explain what other hidden or direct costs the company might have if the party is held on company property or during regular work hours.

2.   Exploration Activity—Comparing Economic Data

The information in the chart below is taken from the U. S. Census Bureau. It compares economic information for three industries in the United States for the years 2007 and 2012.

Type of industry	Annual payroll (in dollars)		Paid Employees	
	2007	2012	2007	2012
Mining	40,687,472,000	59,461,950,000	730,433	848,189
Hospitality	170,826,847,000	196,103,341,000	11,600,751	12,007,689
Health Care and Social Assistance	662,719,938,000	801,239,522,000	16,792,074	18,414,757
Total				

a. For 2007 and 2012, find the total annual payroll (in dollars) and the number of paid employees in the mining, hospitality, and health care/social assistance industries. Record results in the table above. Use the vertical format of the numbers to help you calculate the totals for each column.

b. Now determine the increase in annual payroll from 2007 to 2012 and the increase in the number of paid employees from 2007 to 2012. Complete this part by hand or by using a scientific calculator. Write your results in the table below. *Hint:* The increases are actually differences.

Type of Industry	Increase in Annual Payroll (in dollars)	Increase in Number of Paid Employees
Mining		
Hospitality		
Health Care and Social Assistance		

c. For each type of industry in your table from part *b*, round the increase in annual payroll to the nearest million dollars. Then round your data for the increase in number of paid employees to the nearest hundred thousand. Add two columns to your table from part *b* and enter your results from part *c* in these two new columns.

d. Describe the process that you would use to find the average salary for each industry in 2012. Which data would you use? Determine whether or not your results would be whole numbers.

3.   Conceptual Exercise—Safe Dose

Claudia, a 7-year old girl, weighs 19 kilograms. The manufacturer's recommended safe range for doses of heparin is 10 to 25 units per kilogram per hour. Through a series of steps, you will determine the safe range of doses for Claudia and determine her dose of heparin per hour.

a.   10 units of heparin per kilogram per hour is the low end of the recommended safe dose. What is the lowest safe dose of heparin that should be given to Claudia in one hour?
Hint: The lowest safe dose (number of units) of heparin for Claudia in one hour = the lowest safe number of units of heparin recommended per kilogram per hour × Claudia's weight in kilograms.

Lowest safe dose of heparin for Claudia in one hour: _____ units

b.   25 units of heparin per kilogram per hour is the high end of the recommended safe dose. Calculate the highest safe dose of heparin per hour for Claudia.

Highest safe dose of heparin for Claudia in one hour: _____ units

c.   To summarize the results in parts *a* and *b,* record the recommended safe range (low to high) of units of heparin per hour for Claudia.

From _____ to _____ units of heparin per hour

d. Claudia must receive her dose of heparin as part of an IV or intravenous fluid. The total amount of the IV solution (containing her dose of heparin) is 250 milliliters, and it flows at a rate of 50 milliliters per hour. How many hours will it take to administer 250 milliliters of the intravenous (IV) solution containing heparin? Show your work.

_____ hours

e. Suppose that the total IV solution (250 milliliters) contains 2000 units of heparin. How many units of heparin are administered to Claudia per hour? Show your work.

Hint: The number of units of heparin administered to Claudia per hour = total number of units of heparin in the entire 250-milliliter solution ÷ the number of hours it would take to administer the entire solution (from part *b*).

_____ Units of heparin administered to Claudia per hour

f. Compare your answer in part *e* to the safe range of dosages for Claudia as determined in part *c*. Is this dose within the recommended safe range? Explain why or why not.

4.   Group Activity—Where Did All The Money Go?

Mary Thurmond willed her entire estate to be divided equally among her 22 nieces and nephews. When she died, her estate was valued at over two million dollars. First, in a very simplistic world where there are no fees taken out of the estate, you will determine the amount of inheritance for each of the nieces and nephews. Next, you will determine how to distribute the funds after taxes and probate (the act of verifying the will's validity) fees are subtracted from the estate.

Work with your group members to answer the following questions.

a. In addition to her residence, Mary Thurmond owned a beach house and a farm that was leased to a corporate farmer. Her real estate properties were appraised as shown in the chart below. Calculate and record the total appraised value of her real estate properties in the table.

Property	Appraised Value of Real Estate
Residence	$435,000
Beach House	$350,000
Farm	$730,000
Total	

b. An estate auctioneer estimated the value of her personal belongings at $115,000. Investments in money market accounts and stock market accounts are estimated at $573,000. Find the total estimated value of her personal (not real estate) belongings and investments.

c. What is the total estimated value of all of Mary Thurmond's properties, belongings, and investments? Round this amount to the nearest hundred thousand dollars.

d. Work with your group members to calculate each nephew's and niece's inheritance based on your final rounded estimate in part *c*.

e. An auction was held to sell Mary Thurmond's personal belongings. The real estate properties and stocks were sold, and the money market accounts were closed out. All of this money was held in an escrow account. (Note: An escrow account is money or property placed into the hands of a third party, and is delivered to the grantee only after fulfilling specific conditions.)

The money collected after all real estate, broker, and auctioneer fees were paid is shown in the table below. Find the difference between the actual dollar amount collected after fees and the appraised value. Then determine whether the actual dollar amount after fees reflects a gain or loss in value from the appraised estimate.

Properties	Actual Amount Collected After Fees	Difference between Actual Amount and Appraised Value	Gain or Loss in Value?
Residence	$394,190		
Beach house	$360,100		
Farm	$522,000		
Personal belongings	$53,250		
Stocks and Money Market	$427,930		
Total			

What is the total amount of money that was deposited into the escrow account?

_____

f. Calculate the share of the "total in escrow" for each nephew and niece. Find the difference between this share and the estimated share based on appraised values (from part *d*).

g. Probate court fees came to $55,000, and taxes totaled $575,230. What total sum was left to divide among the heirs after these costs are paid from the escrow account?

h. After taxes and probate fees are paid, how much money will finally be given to each nephew and niece? Round to the nearest cent.

i. Calculate the difference between what each niece and nephew might have expected given the appraised values (from part *d*) and their actual share (from part *h*).

1.   **Extension Exercise—Temperature Scales**

If C is the temperature in degrees Celsius, and F is the temperature in degrees Fahrenheit, then we can represent the relationship between temperature scales with the following formula.

$$C = \frac{5}{9} \cdot (F - 32)$$

a.   Convert the Fahrenheit temperatures of 32°F, 23°F, and 14°F to the Celsius scale. Use the two-step process described next to arrive at your answer.

    Step 1: Subtract 32 from the given temperature.

    Step 2: Multiply by $\frac{5}{9}$.

$$C = \frac{5}{9} \cdot (32 - 32) \qquad\qquad C = \frac{5}{9} \cdot (23 - 32) \qquad\qquad C = \frac{5}{9} \cdot (14 - 32)$$

$$C = \qquad\qquad\qquad\qquad\quad C = \qquad\qquad\qquad\qquad\quad C =$$

b.   Let's reverse the above procedure by first describing the opposite of Step 2 and then the opposite of Step 1. This new two-step procedure will allow you to convert any Celsius temperature to Fahrenheit temperature.

    Reverse of Step 2: _____

    _____

    Reverse of Step 1: _____

    _____

c.   Show how to convert the Celsius temperatures of 0°C, –5°C, and –10°C to Fahrenheit temperatures using the reverse procedure developed in part *b*.

d. Letting *C* represent the temperature in degrees Celsius and *F* represent the temperature in degrees Fahrenheit, suggest a formula for converting from degrees Celsius to degrees Fahrenheit.

e. Temperatures in space are measured in degrees Kelvin. Find the relationship between the Kelvin scale and the Celsius scale by observing the last two columns.

Temperature of	Kelvin	Celsius
Earth's Sun	6000	5727
Water Boiling	373	100
Water Freezing	273	0
Absolute Zero	0	–273

- What operation can you perform to convert a Kelvin temperature to a Celsius temperature?

- What operation can you perform to convert a Celsius temperature to a Kelvin temperature?

f. If *K* is a temperature in degrees Kelvin and *C* is a temperature in degrees Celsius, then suggest a formula for converting degrees Kelvin to degrees Celsius and a formula for converting degrees Celsius to degrees Kelvin.

2.   Exploration Activity—Balancing Your Account

You just received your bank statement, which shows an end-of-month balance of $25 in your checking account. Where did all the money go? You decide to compare last month's deposits and withdrawals. A good way to do this is to take the balance from the start of last month. Then add a positive number for deposits, and add a negative number for withdrawals. The following table shows this process, and the first calculation is completed for you.

Transaction	Withdrawal (−) ($)	Deposit (+)	Calculation	Balance ($)
Start of Month				+625
Rent	−550		+625 + (−550)	+75
Car Loan	−250			
Hair Cut	−25			
Paycheck		+600		
Auto Repair	−575			
Electric Bill	−75			
Telephone Bill	−110			
Cable	−25			
Paycheck		+600		
Groceries	−190			
End of Month				+25

(↑ for part *d*)   (↑ for part *e*)

a. Finish filling in the table by showing the calculation needed to obtain the balance after each transaction. Place the result of each calculation in the balance column as either a positive or negative number. Complete all calculations by hand, and then go back to check your work with a calculator.

b. Observe all the calculations where you added a positive number and a negative number.

- If you temporarily ignore the signs of the numbers, then what operation did you actually perform on the two numbers in order to obtain the correct balance?

- After performing the operation, what about the two numbers determined whether the balance was a positive number or a negative number?

- In your own words write a general procedure for adding positive and negative numbers.

c. Now observe the calculations where you added two numbers with the same sign (either both positive or both negative).

- If you temporarily ignore the signs of the numbers, then what operation did you actually perform on the two numbers?

- After performing the operation, what about the two numbers determined whether the balance was positive or negative?

- In your own words write a general procedure for adding two numbers with the same sign.

d. Return to the table and add up all the withdrawals in the second column. Record the sum below and in the table (*End of Month* row, *Withdrawal* column).

e. Return to the table and add up all the deposits in the third column. Record the sum below and in the table (*End of Month* row, *Deposit* column).

f. Analyze the results of parts *d* and *e*. Then explain what is happening to your cash flow over this one-month period.

g. How did you manage to end up with +25 dollars when more money was going out than coming in?

h. Assuming you have no auto repairs or hair cuts next month, estimate your ending balance for next month. Show all your work.

3.   **Conceptual Exercise—Rent-a-Calculator**

Suppose you decide to start a business renting graphing calculators to other students at your college. Before you can begin your business, you need to purchase some graphing calculators. You have enough money saved to buy 50 reconditioned calculators from a wholesaler for $45 per calculator. On campus, you advertise that any student can rent a calculator for one semester at a fee of $15 per semester, but it must be returned in good condition. You plan on running this business over a two-year period that will cover six semesters (fall, spring, and summer of each year).

a.   What is the initial *cost* to start your business? Show your work.

b.   If you manage to rent all 50 calculators for two consecutive semesters, what will be your *revenue* at the end of this time period? Revenue is the money coming in from the rentals. Show your work.

c.   Use your results from parts *a* and *b* to calculate the *profit* after two semesters. Show your work.

$$Profit = Revenue - Cost$$

d.   If the *profit* is negative, then explain the status of the business in terms of revenue and cost. Use complete sentences in your explanation.

e.   How many calculators must you rent during your third semester of business to break even? Show all work.

f. By the end of the third semester, your business does break even. Now you set the goal of making $3000 profit over the next three semesters. To accomplish your goal, what is the new minimum rental price that must be charged for the calculators per semester?

g. What type of risk are you taking by charging a price greater than $15? Explain using complete sentences.

h. Suppose the increase in price causes a decrease in demand such that you are only able to rent 40 calculators in each of the last three semesters of running your business. How much are you short of your $3000 goal if you use the rental price from part *f* ?

4.   Group Activity—Elevator Talk

Suppose you work in a building that has three floors above ground level and two floors below ground level. Imagine that the figure below represents an elevator that takes people up and down to different floors. Assume that the distance between successive floors is 10 feet.

Work with your group members to answer the following questions.

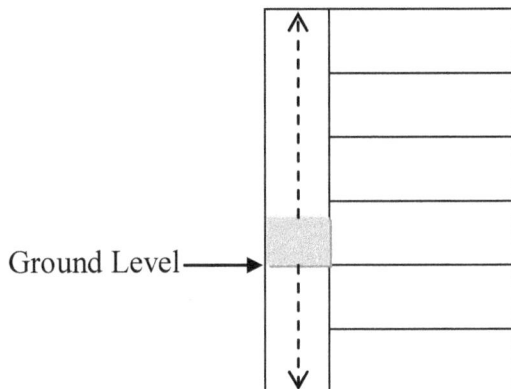

Ground Level

a. Label the ground level floor as 0, and then label the floors above floor 0 with the positive integers 1, 2, 3, and 4 and below floor 0 with the negative integers −1 and −2.

b. Sharlitha enters the elevator at ground level, travels up to floor 1 to drop off a letter, and then goes down to floor −2 to pick up some supplies. To the right of the figure above, draw and label two arrows pointing up or down to visualize how positive and negative numbers can represent her vertical movement on the elevator.

c. Use addition of positive and negative integers to show the calculation that represents Sharlitha's elevator trip and the solution that represents the floor she ends up on (floor −2).

d. Let 10 represent the number of feet Sharlitha traveled up and −30 the number of feet she traveled down. Use subtraction to represent the total distance that Sharlitha traveled between floors.

e. After picking up supplies on floor –2, Sharlitha gets back on the elevator with two other people, Terry and Marco. Marco says to Terry, "I do not understand how these floors are numbered. If zero means *nothing*, then it's impossible for anything to be less than nothing." State whether you agree or disagree with Marco and support your answer in complete sentences.

f. Terry gets out on floor –1. Then Marco starts up a conversation with Sharlitha saying, "Oh, I have the money I borrowed from you yesterday at lunch." Sharlitha responds, "Happy birthday! Consider that money as my treat for lunch."

If *b* represents the amount of money that Marco borrowed from Sharlitha, then how would you interpret the following rule? $-(-b) = b$ $(b > 0)$
Hint: The great mathematician Leonard Euler once said, "…*to cancel a debt signifies the same as giving a gift.*"

g. Marco and Sharlitha leave the elevator at floor 0. Before going back to work, Marco describes his five-month battle trying to lose weight. "In the first month of dieting, I cheated all the time and gained 10 pounds. During the second month, I lost 5 pounds. Then, in the third month, I lost 7 pounds. During the fourth month, I lost 3 pounds, and now I have gained 4 pounds during the fifth month." Marco refers to his weight journal below:

Month	January	February	March	April	May
Weight Change	+10	–5	–7	–3	+4

- Show the calculation needed to find the total change in weight over the 5-month period.

- How much weight did Marco gain or lose in total? Express your answer as a positive or negative integer.

1.   Extension Exercise—Mystery Dose

It is difficult to calculate the exact surface area of a human body since it is not flat or shaped like a box. However, health professionals have come up with a formula to approximate the number of square meters it would take to cover a human body. All you need to determine a person's body surface area (BSA) is that person's height and weight. Once calculated, a person's BSA can be used to determine the proper individual dose for some medications.

The table below contains data on two people. In each case, either the BSA or the proper dose is unknown and has been replaced by a variable, *x* or *y*.

Person	Body Surface Area (square meters)	Recommended Rule for Finding the Proper Dose	Proper Dose (milligrams)
Linda	1.5	3.7 mg. per BSA of 1 sq. meter	*x*
Jim	*y*	9.3 mg. per BSA of 1 sq. meter	23.3

a. For Linda, label what *x* represents in words by completing the "Let *x* =" statement below. Then write a general formula for Linda by using the known values to find the value for *x*. Round results to the nearest tenth of a milligram. Show your work below.

Let *x* = _____

General formula for Linda: *x* = _____

b. For Jim, label what *y* represents in words by completing the "Let *y* =" statement below. Then write a general formula for Jim by using the known values. If needed, solve for *y*. Round results to the nearest tenth. Show your work below.

Let *y* = _____

General formula for Jim: *y* = _____

c. For the next five days, Linda's dose needs to slowly increase. Use Linda's BSA from the table on the previous page and the table below to find the proper dose (in milligrams) for Linda on each day. Round results to the nearest tenth of a milligram. Show calculations and record answers in the table below.

Milligrams per 1 square meter of BSA		Calculations for Linda	Amount of Dose for Linda (milligrams)
First day:	3.7		
Second day:	5.5		
Third day:	7.4		
Fourth day:	9.3		
Fifth day:	11.1		

d. Jim's dose each day needs to slowly decrease for the next five days. Use Jim's BSA from your answer in part *b* and the table below to find the proper dose (in milligrams) for Jim on each day. Round results to the nearest tenth of a milligram. Show calculations and record answers in the table below.

Milligrams per 1 square meter of BSA		Calculations for Jim	Amount of Dose for Jim (milligrams)
First day:	9.3		
Second day:	8.6		
Third day:	7.5		
Fourth day:	5.4		
Fifth day:	3.5		

2.   Exploration Activity—Community Fund-Raiser

Your community is hosting a fund-raiser for charity. Decorations, pamphlets, and other miscellaneous items have been donated. Volunteers have planned the event and will prepare the food, so there are no labor costs for the event. It will cost about $5.25 per person to buy the food, and $320 to rent a hall for the evening.

a. Do the necessary calculations and fill in the following table to show how much the event will cost depending upon the number of people attending in the first column.

No. of People Attending	Cost of Hall Rental ($)	Cost of Food, $5.25 per person ($)	Total Cost ($)
50			
100			
150			
200			
250			
300			
350			

b. Write an expression that represents the cost of food if *n* people attend the fund-raiser.

cost of food for *n* people: _____

c. Write an expression that represents the total cost if *n* people attend the fund-raiser.

total cost for *n* people: _____

d. If C represents total cost and n represents the number of people attending the fund-raiser, write an equation that relates C and n. *Hint:* Use your expression from part *c*.

equation: _____

e. If people are charged $15 each for the dinner, calculate the total revenue for the dinner. Then copy the total cost from the table in part *a* into the 3rd column, and calculate the profit, which is revenue minus total cost.

No. of People Attending	Total Revenue ($) (at $15 per person)	Total Cost ($) (from part *a*)	Profit ($) = Revenue – Cost
50			
100			
150			
200			
250			
300			
350			

f. If R represents revenue and n represents the number of people attending the fund-raiser, then write an equation for R in terms of n.

$R =$ _____

g. Record your equations from parts *f* and *d* in the spaces provided below.

$R =$ _____ $C =$ _____

h. The profit of the fund-raiser will be the revenue minus the cost. Use the formulas in part *g* to write an expression for profit P. Then simplify the expression by removing parentheses and combining like terms.

$P =$ _____ $-$ (_____) $=$ _____

          Revenue   Cost   Simplified

i. Use the equation for profit in part *h* to determine how many people would have to attend the event to make a profit P of $800. Round your answer to the nearest whole person.

3.   Conceptual Exercise—Money in Trees

When determining the dollar value of a tree, you must estimate the size of the tree. To do this, you need to calculate the tree's **d**iameter **b**reast **h**eight (**dbh**), which is the diameter of the tree trunk at 4 ½ feet above the ground (breast-high on a man of average height).

a. There is no real monetary value of a tree until it obtains a diameter breast height (dbh) of about 14 inches. Use the formula, $C = \pi d$, to find the minimum circumference C (at breast height) for a tree to begin to have real monetary value. Round your answer to the nearest whole inch. *Hint:* $d = 14, \pi \approx 3.14$

Minimum circumference: $C =$ _____ inches

b. For each circumference given below, find the diameter breast height of the tree. First, substitute the given value for C into the equation, $C = \pi d$. Then solve the equation for d. Show your substitution and work in the second column, and record your answer for d in the last column. Round d to the nearest inch. *Hint:* $\pi \approx 3.14$

Circumference at Breast Height (inches)	Equation (show work)	What is d ? (rounded to nearest inch)
47		
53		
66		
88		

c. If you know the circumference of a tree, how would you solve the equation $C = \pi d$ for d (so that it begins as $d = $?) To check your equation, use it to calculate the diameter breast height d for each circumference in part b and see if you get the same answer.

$$d = \underline{\hspace{5cm}}$$

d. If a tree's diameter decreases 1.5 inches every 12 feet up the tree (on average), then how much will the tree decrease in diameter for each 1 foot increase in height? *Hint:* You can set up and solve a proportion to figure this out.

Decrease in diameter per foot = _____ inches

e. Using the answer from part d, explain how you would find the decrease in diameter of a tree at 20 feet.

f. Generalize the process that you used in part e and write a formula for the decrease in the diameter over x feet.

Decrease in diameter = _____

4.   Group Activity—Springs in the Ozarks

Mark, a geologist, recently traveled to the Ozarks in Missouri and Arkansas. He is fascinated by the constant flow of underground springs that feed the many clear, deep pools in the area, and thus, he gathered the following information about two of these underground springs: Round Springs and Big Springs, both in Southern Missouri.

Name of Spring	Average Amount of Water Flow Per Day (millions of gallons)	Maximum Water Flow Ever Measured Per Day (millions of gallons)
Big Spring	276	800
Round Spring	26	300

The formula used to calculate the flow rate of a spring is $A = r \cdot t$, where A is the amount of water flow, r is the rate of water flow, and t is the measured amount of time. For the data in this table, t equals one day. Remember that when using formulas, the units must match.

Work with your group members to answer the following questions.

a.   Solve the formula $A = r \cdot t$ for r.

b.   Rates are measured in such units as miles per hour, feet per second, gallons per minute, and kilograms per hour. At the springs, what unit is used to measure the rate of water flow?

c.   What is the average rate of water flow r for Big Springs? Use the data in the 2nd column of the table and the formula for r from part *a*.

       Average rate of water flow, Big Springs: $r =$ _____ millions of gallons of water per day

d.   What is the average rate of water flow r for Round Springs? Use the data in the 2nd column of the table and the formula for r from part *a*.

       Average rate of water flow, Round Springs: $r =$ _____ millions of gallons of water per day

e. What is the maximum rate of water flow r for Big Springs? Use the data in the 3rd column of the table and the formula for r from part a.

Maximum rate of water flow, Big Springs: r = _____ millions of gallons of water per day

f. What is the maximum rate of water flow r for Round Springs? Use the data in the 3rd column of the table and the formula for r from part a.

Maximum rate of water flow, Round Springs: r = ___ millions of gallons of water per day

g. Calculate the estimated amount of water to flow from each of the springs in an *average* year. Give your answers in millions of gallons per year. (1 year = 365 days)

Amount of water through Big Springs: _____ millions of gallons per year

Amount of water through Round Springs: _____ millions of gallons per year

h. About how many days would it take for one billion gallons of water to flow through each of the springs using the average amount of flow per day? Round your answers to the nearest whole day. *Hint:* Convert one billion into millions of gallons. Then substitute the result into the equation for A, and solve for t.

Big Springs will have one billion gallons of water flow through it in _____ days.

Round Springs will have one billion gallons of water flow through it in _____ days.

1.   Extension Exercise—Monitoring Your Stocks

Suppose you go online at 12:29 p.m. to check the prices per share of stock that you own in two companies. First you find the following data on Dover Computer Corporation.

Dover Computer Corporation

Ticker Symbol	Today's High	Today's Low	Last Price	Prior Close	Change	Trade Time
Dover	$42\dfrac{15}{16}$	$39\dfrac{1}{2}$	$41\dfrac{3}{8}$	$43\dfrac{9}{16}$	$\downarrow 2\dfrac{3}{16}$	12:29 p.m.

a.   As of 12:29 p.m. this trading day, what is the difference between today's high price per share and today's low price per share? Show all your work.

b.   The Last Price represents the price of a share as of Trade Time 12:29 p.m., and the Prior Close was the price per share at the close of trading on the previous day. Show how subtracting the Last Price from the Prior Close price yields the difference of $2\dfrac{3}{16}$. Then explain what the arrow indicates in the "Change" column.

c. If you own 1000 shares of Dover, then how much value did your Dover investment lose
during the time from the Prior Close price until the Trade Time of 12:29 p.m. today? Do you
think this is a cause for concern? Explain.

d. Now observe data from Medicon Corporation, another company that you own stock in. What
must the "Last Price" be in order to obtain the "Change" of $\uparrow 2\frac{19}{32}$?

Medicon Corporation

Ticker Symbol	Last Price	Prior Close	Change	Trade Time
Medicon	?	$5\frac{1}{4}$	$\uparrow 2\frac{19}{32}$	12:29 p.m.

e. If you own 9600 shares of Medicon, then how much value did your Medicon investment gain
during the time between the Prior Close price and the Trade Time of 12:29 p.m. today? Do
you think this is a cause for celebration? Explain.

2.   **Exploration Activity—Bills and Coins**

In the United States, the value of a coin or bill is based on a fraction or multiple of the basic unit, the dollar. A half-dollar and a quarter are so named because they are ½ of a dollar and ¼ of a dollar, respectively.

a.   In the table below, write how many coins it takes to equal one dollar and what fraction of a dollar (or 100 cents) each coin is. The first row has been completed for you.

Coin	Number of Coins Needed to Make $1	One Coin is What Fraction of a Dollar?	Simplify the Fraction.
Penny	100	$\dfrac{1}{100}$	$\dfrac{1}{100}$
Nickel			
Dime			
Quarter			
Half-dollar			

b.   Three quarters = _____ pennies

c.   Write the fraction of a dollar that represents 75 pennies. Now write the fraction of a dollar that represents three quarters. Show that the two fractions are equal using cross products.

d.   One half-dollar has the same value as how many pennies? nickels? dimes? quarters?

1 half-dollar = _____ pennies = _____ nickels = _____ dimes = _____ quarters

Now represent this series of equal values as equivalent fractions. The first fraction is written for you. Prove that all are equivalent fractions by finding the cross products of the first two fractions, the cross products of the second and third fractions, and so on.

$$\frac{1}{2} =$$

Vanessa works on the floor of a casino selling tokens and chips to customers. The tokens come in rolls, and the rolls are put in *squares* of 25 rolls. A *bucket* is equal to two squares.

e. Complete the table below to determine the value of one roll of each type of token, the value of one square of tokens, and the value of one bucket of tokens. The first row has been completed for you.

Type of Token	Value of Token ($)	No. of Tokens in a Roll	Value of one Roll ($)	Value of one Square (25 rolls) ($)	Value of one Bucket (2 squares) ($)
H	$\frac{1}{2}$	40	$20	$500	$1000
Q	$\frac{1}{4}$	40			
T	$\frac{1}{10}$	100			
N	$\frac{1}{20}$	100			

f. Vanessa invested her hard-earned money and bought 480 shares of stock at $7\frac{3}{8}$ dollars per share. How much did the 480 shares cost? Show your work.

g. In the saying, "Two bits, four bits, six bits, a dollar," follow the pattern to determine how many bits are in one dollar and the value of 2 bits, 4 bits, 6 bits, and 8 bits as a fraction of a dollar. How many quarters equal 2, 4, 6, and 8 bits?

Number of bits in a dollar: _____

No. of Bits	Fraction of a Dollar	×	Number of Quarters in a Dollar	Number of Quarters Equal to 2, 4, 6, or 8 Bits
2		×	4	
4		×	4	
6		×	4	
8		×	4	

3.   Conceptual Exercise—Team Effort

Doug spends three hours every Saturday morning mowing the lawn while his son, Michael, watches cartoons on TV. The last time that Michael tried to mow the lawn, it took him three hours to complete half the job. Still, Doug figures that if he borrows his neighbor's lawn mower and he and his son work together, this will give his son a sense of accomplishment and a positive work ethic.

a.   Since Doug is able to mow the whole lawn in 3 hours, what fraction of the lawn can he mow in 1 hour?

b.   If Michael can mow half the lawn in 3 hours, then how long will it take him to mow the whole lawn alone?

c.   Using your result from part *b,* what fraction of the lawn can Michael mow in 1 hour?

d.   Working together, what fraction of the lawn can Doug and Michael mow in 1 hour? Show all work including the process of finding a least common denominator. Write your answer in simplest form.

e.   How long does it take to mow the whole lawn with father and son working together?

f.   Why is it helpful to find the fraction of the lawn each person could mow in 1 hour?

g. The concept of finding a common denominator can be visualized with a geometric model. Let the whole lawn be represented by the rectangle below. Split the rectangle into the appropriate number of parts to represent how long it takes Doug to mow the entire lawn. Then shade in the portion of the rectangle to show the part of the job that Doug completes in 1 hour working alone.

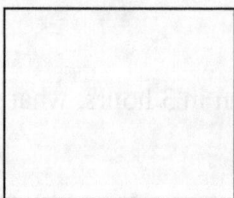

h. Now, let the same lawn be represented by another large rectangle given below. Split the rectangle into the appropriate number of parts to represent how long it takes Michael to mow the entire lawn. Shade in the portion of the rectangle to show the part of the job that Michael completes in 1 hour working alone.

i. If Doug and Michael work together, then you can add the fractional amount that each mows in 1 hour separately. Show this geometrically by placing your two models from parts *g* and *h* below and find their sum.

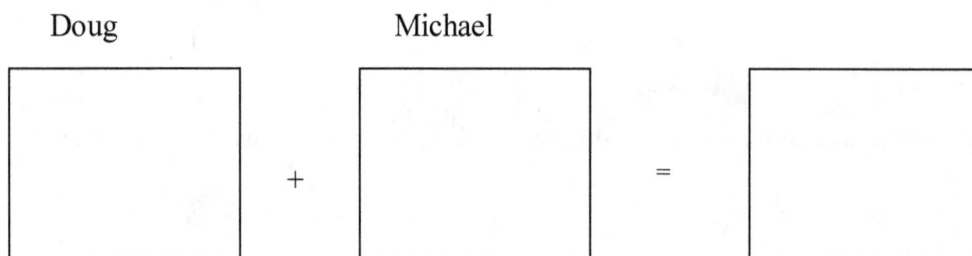

Doug   Michael

+   =

j. What do your answers in part *i* and part *d* have in common?

4.   **Group Activity—Cubes and Fractions**

<u>Materials Needed:</u>
- Paper to make 2 squares--one with a side length of ½ inch to represent a ½-inch cube, and the other with a side length of ¾ inch to represent a ¾-inch cube
- ruler

A **cube** is a three-dimensional object shaped like a sugar cube where length, width, and height are the same measurement. A cube that is ½ in. x ½ in. x ½ in. is called a half-inch cube (½-inch cube). A cube that is ¾ in. x ¾ in. x ¾ in. is called a three-quarter-inch cube (¾-inch cube). You and your group members will use paper squares to represent one side of a ½-inch cube and one side of a ¾-inch cube.

Work with your group members to answer the following questions.

a. If you line up 13 half-inch-cubes side by side, as illustrated below, then what is the total length of the cubes together? Use a ruler and your ½-inch paper cube (square) to measure the length of 13 half-inches.

Length of 13 half-inch cubes _____ inches

b. How would you use multiplication to find the total length of 13 half-inch cubes? Express the length as an improper fraction and as a mixed number. Show your work.

_____ inches   =   _____ inches
(improper fraction)   (mixed number)

c. If you line up 16 three-quarter-inch cubes side by side, then what is the total length of the cubes together? Use a ruler and your ¾-inch paper cube (square) to measure the length of 16 three-quarter-inches.

d. How would you use multiplication to find the total length of 16 three-quarter-inch cubes? Show your answer as an improper fraction, and then write your answer in simplest form.

Improper fraction: _____ inches   **Simplest form:** _____ inches

How might you build a box with ½-inch cubes? One example is shown below to the right. The bottom layer of the box is shown below to the left.

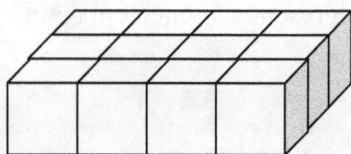

The bottom layer of the box has 3 rows of half-inch cubes and 4 columns or 4 cubes in each row.

One half-inch cube

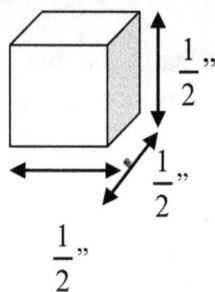

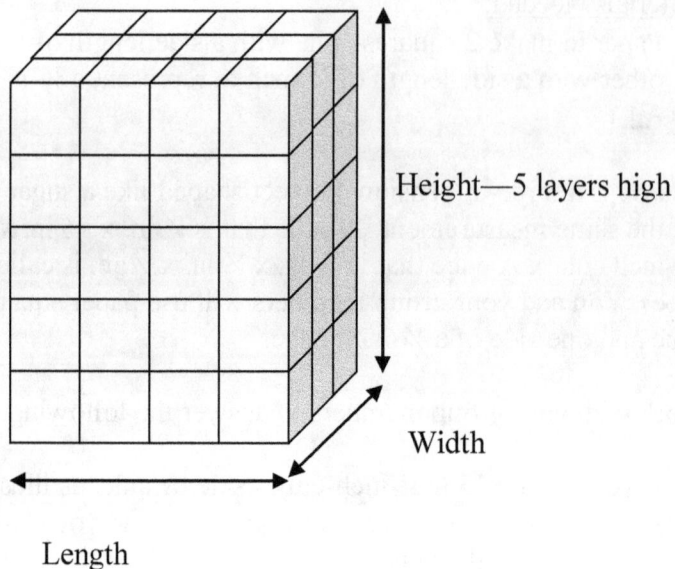

Height—5 layers high

Width

Length

e. Let's now find the length, width, and height of the five-layered box. Use the illustration above to help you find the answer. Express each measurement in terms of number of cubes and number of inches. If a measurement does not result in a whole number of inches, then express your answer as an improper fraction and as a mixed number.

Length: _____ cubes = _____ inches

Width: _____ cubes = _____ inches

Height: _____ cubes = _____ inches

f. The volume of a rectangular box is found by multiplying length × width × height of the box. Calculate the volume of the 5-layered box above using the inch measurements that you found in part *e*. Write the volume in simplest form. If the result is not a whole number, then express your answer as an improper fraction and as a mixed number.

Volume = _____ x _____ x _____ = _____ cubic inches
           length   width   height

1.   Extension Exercise—A Year at MIU

Suppose you have just started as an exchange student at Maritime International University (MIU). The apartment where you will live is 10 kilometers from campus.

a. If you drive your car from your apartment to the MIU campus, how does travel time depend on average speed (rate)? To find out, complete the table below. Write hours in simplest fractional form. Then give minutes in whole number or decimal form rounded to the tenths place.

$$\text{Time} = \frac{\text{Distance}}{\text{Rate}}$$

Distance (km)	Rate (km/hr)	Time (hours)	Time (minutes)
10	10	1	60
10	20		
10	36		
10	40		
10	45		
10	55		
10	60		
10	70		
10	80		

b. Examine the second and third columns. Every time you double the rate, what happens to the trip time? Answer in complete sentences using examples to support your conclusion.

c. Suppose your average walking pace is 2.7 kilometers per hour. How long will it take you to walk from your apartment to the MIU campus? Show all work and present your answer, the time, in decimal form. Then convert this time to minutes. Round all times to the nearest tenths place.

The fuel economy of your car is 25.6 kilometers per gallon when traveling at a rate of 40 kilometers per hour. Your car's gas tank holds 16.5 gallons.

d. If you are driving at a constant rate of 40 kilometers per hour, what is the total distance (in kilometers) that your car can travel on one tank of gas if all the gas is used?

e. Suppose the price of gasoline is equivalent to $1.99 per gallon. What is the cost of filling an empty tank? Round your answer to the nearest cent.

f. What is the cost of gas per kilometer if your speed is 40 kilometers per hour? Round your answer to the nearest cent.

g. If your first class is at 8 a.m. and you drive at an average speed of 40 kilometers per hour, then at what time should you leave for school each morning? Explain why there might be more than one good answer to this question.

h. If you drive your car from your apartment to the MIU campus each day with no side trips, then how often should you fill your car's gas tank? Round your results to the nearest whole number. *Hint:* You will use your answer from part *d.*

i. If the cost of gasoline stays fixed at $1.99 per gallon for the entire 15-week semester, then about how much will fuel cost over the entire semester? Assume that you do not drive on weekends and that classes occur during the 5-day week. Round results to the nearest cent.

2.   **Exploration Activity—Exchange Rates**

In October 2018, the rates of exchange between the United States dollar (USD) and selected foreign currencies were as listed in the table below. Exchange rates can tell you what U.S. dollars are equivalent to in a foreign currency or what a foreign currency is equivalent to in U.S. dollars. For example, on October 17, 2018, 1 USD equaled nearly 7 Chinese yuan (6.93 CNY), almost 19 Mexican pesos (18.81 MXN), or about 112 Japanese yen (112.19 JPY). The values in the table list the exchange rates as units of foreign currency per 1 USD.

The euro/dollar exchange rate can be expressed as $\dfrac{0.87\text{ EUR}}{1\text{ USD}}$, meaning that there are 0.87 euro to every 1 U.S. dollar. This rate can also be written as $\dfrac{1\text{ USD}}{0.87\text{ EUR}}$, meaning that there is 1 U.S. dollar to every 0.87 euro. Assume that all transactions referred to in this activity occur during October 2018 at the rates shown in the table below.

Foreign Currency	Rate Units per 1 USD	Foreign Currency Price of a $500 (USD) TV	Rate USD / unit
Australian dollar - AUD ($)	1.40		
British pound - GBP (£)	0.76		
Canadian dollar - CAD ($)	1.30		
Chinese yuan - CNY (¥)	6.93		
Euro - EUR (€)	0.87		
Japanese yen - JPY (¥)	112.19		
Mexican peso - MXN ($)	18.81		

(Exchange rates recorded on October 17, 2018)

a. Suppose you buy a basic TV for $500 in the U.S. The second column of the table expresses the number of foreign currency units that are equivalent to 1 U.S. dollar (that is, units per 1 USD). Complete the third column of the table by converting the price of the same TV to the currency of each country.

b. If a Canadian citizen buys a car in the U.S. that sells for 20,000 USD, how many Canadian dollars (CAD) must be exchanged to make the purchase in U.S. dollars? Show all work.

c. Suppose a U.S. worker would like to take a vacation to Mexico and decides to set aside one week's pay for spending money while she is on vacation. She earns 21.42 USD per hour (after all payroll deductions) and works a 40-hour week. If she exchanges her entire week's earnings for Mexican pesos, how many pesos will she be able to take with her on vacation? Use the exchange rate given in the table, show all your work, and round your answer to the nearest hundredth.

d. Use the fourth column in the table to list each of the exchange rates as U.S. dollars per unit of foreign currency. For example, 1 USD per 0.87 EUR can be expressed as the ratio

$$\frac{1\,\text{USD}}{0.87\,\text{EUR}} \approx 1.149 \text{ dollars/euro},$$ meaning there are 1.149 U.S. dollars in 1 euro, or that 1 euro is approximately equal to 1.149 U.S. dollars. Use a calculator as needed. Round results to the nearest thousandth.

e. If you took a vacation on the French Riviera that cost 27,672.30 EUR, how many U. S. dollars would need to be exchanged for the trip? Round your answer to two decimal places.

f. Suppose you took a job in Japan that has a salary of 7,225,800 yen. What would be your salary in U.S. dollars? Round to the nearest dollar.

The table on the first page of this activity is set up for the conversion between U.S. dollars and various foreign currencies. However, you can also calculate the exchange rate of each foreign currency in terms of any other foreign currency. For example, we know that 1 U.S. dollar equals 0.87 EUR, and 1 U.S. dollar equals 0.76 GBP. This implies that 0.87 EUR = 0.76 GBP.

This is an example of cross-exchange rates. The EUR/GBP exchange rate can be calculated by dividing both sides of the equality by 0.76,

$$0.87 \text{ EUR} = 0.76 \text{ GBP}$$
$$\frac{0.87 \text{ EUR}}{0.76} = \frac{0.76 \text{ GBP}}{0.76}$$
$$1.145 \text{ EUR} \approx 1 \text{ GBP}$$

This tells us that 1 British pound (GBP) is approximately equal to 1.145 euros (EUR).

g. Use the ideas developed above to complete the following table of cross-exchange rates. Round to the nearest thousandth.

Currency	units per 1 USD	units per 1 AUD	units per 1 GBP	units per 1 CAD	units per 1 CNY	units per 1 EUR	units per 1 JPY	units per 1 MXN
USD	1							
AUD	1.40	1						
GBP	0.76		1					
CAD	1.30			1				
CNY	6.93				1			
EUR	0.87					1		
JPY	112.19						1	
MXN	18.81							1

3.   Conceptual Exercise—Traveling for Big Brother

You have been asked to be a consultant for the Big Brother Oil Corporation's mining and coal division. Your task is to examine the most economical way to transport coal to an electric company.

a. Suppose preliminary research shows three possible methods of transportation:

- Trucks traveling 150 miles to the destination
- Freight trains covering 800 miles of railroad to reach the destination
- A barge going for 200 miles on water, then a freight train covering 700 miles of railroad

Do you have enough information to make a recommendation to Big Brother? If yes, go to the next page and write a brief report on the best course of action. Otherwise, state what additional things must be known to make an informed decision.

b. More research reveals the following data on the cost of transporting coal.

- The truck costs $0.17 per ton per mile.
- The freight train costs $0.03 per ton per mile.
- The barge costs $0.02 per ton per mile.

Based on the information gathered so far, which method would you recommend to minimize the cost of moving the coal? Show all work below, then in part *d* write a report to Big Brother Oil Corporation justifying your conclusions.

c. Use the data shown to answer the question below the data.

- The train travels at 70 miles per hour.
- The barge travels at 25 miles per hour.
- The truck travels at 55 miles per hour.
- The train conductor makes 3 stops for a ½ hour break at each stop.
- The trucker makes 3 stops of 15 minutes each.
- The barge conductor travels straight through.
- Time = $\dfrac{\text{distance}}{\text{speed}}$

Assuming that all of the data is true, which of the routes listed in part *b* will take the least amount of time? Show your work, and report the times in hours and minutes.

d. Write a brief report to Big Brother Oil Corporation explaining the most economical way to transport coal to the electric company. Is this the best choice for time reasons? Use your answers from parts *b* and *c* to explain your choice. Give both the pros and cons of your decision.

4.   Group Activity —The Donut Dilemma

The alarm jolts you awake on a Monday morning. It is your turn to bring donut holes to your 8:00 a.m. math class. The local donut shop sells donut holes in boxes of 20, 45, and 60. You have $10 of spending money that must last until Friday, and you are wondering what to purchase at the donut shop. Work with your group members to discuss and answer each of the questions.

a. List what questions you need to consider before you buy any donut holes.

b. Suppose the following data is known:

- One box of 20 donut holes costs $1.99 including tax.
- One box of 45 donut holes costs $3.29 including tax.
- One box of 60 donut holes costs $3.69 including tax.
- There are 33 students registered for your math class.
- On average, 5 students are absent from any class meeting.
- On average, each student eats 3 donut holes.

How many boxes of each size should you buy to make sure that everyone has enough to eat, while at the same time minimizing the cost? Show all work and justify your answer in words using complete sentences.

c. Explain why part *b* has more than one right answer.

d. What is the maximum number of donut holes that can be bought for $7.00? Show all work and justify your answer in words using complete sentences.

e. How many correct answers are there to part *d*? In what way is this question different from part *b*? Explain.

1.   Extension Exercise—Batting Champ

Going into the last day of the major league season, Babe Boggs and Kirby Lockett had identical batting records of 200 hits in 600 at bats. On the last day of the season, Boggs had 7 hits in 8 times at bat, while Lockett had 9 hits in 12 times at bat. Who do you think won the batting title? *Note:* The batting title goes to the hitter with the highest ratio of "hits" to "at bats."

a.   Calculate the identical batting average of Boggs and Lockett just before the last day of the season. The batting average is the ratio (fraction) of "hits" to "at bats." Express this ratio as a fraction in lowest terms, a decimal rounded to the thousandths place, and a percent.

$$\frac{hits}{at\ bats} =$$

b.   What is the practical meaning of the batting average when expressed as a percent?

c.   Calculate Boggs' and Lockett's batting average just for the last day of the season. Express the ratio as a fraction in lowest terms, a decimal rounded to the thousandths place, and a percent. Who had the most hits on the last day? Who had the higher batting average if we just use the last day statistics?

Last Day Statistics

Boggs: Lockett:

Most Hits:

Highest Average:

d.   Which player do you think won the batting title? Explain using complete sentences.

e. Calculate the final batting average for each player at the end of the baseball season. Express the ratio as a fraction in lowest terms, a decimal rounded to the thousandths place, and a percent. Who had the most hits at the end of the season? Who had the higher batting average at the end of the season?

<u>Final Season Statistics</u>

Boggs: $\dfrac{hits}{at\ bats} =$ Lockett: $\dfrac{hits}{at\ bats} =$

Most Hits:

Highest Average:

f. Who won the batting title? Explain using complete sentences.

g. Let's return to the last day of the season when Boggs had 7 hits in 8 at bats, while Lockett had 9 hits in 12 at bats. The table below shows how Lockett progressed through the last game. Fill in the last three rows, giving his season batting average after each "at bat" as a ratio, a decimal, and a percent. Round the decimal to the thousandths place.

Time at Bat	1st	2nd	3rd	4th	5th	6th	7th	8th	9th	10th	11th	12th
Outcome	Hit	Hit	Hit	Out	Hit	Hit	Hit	Hit	Out	Hit	Out	Hit
Current Average (ratio)												
Current Average (decimal)												
Current Average (percent)												

2.   Exploration Activity—A Diverse India

In 2017, India celebrated its 70th anniversary of independence. The country was created from 600 states into a diverse nation with many ethnic groups that speak 18 official languages and practice 7 major religions. India is a study in contrasts when it comes to food, education, and wealth. Although the country is agriculturally self-sufficient, nearly 30% of children under 5 are malnourished. The literacy rate jumped from 12% in 1947 at the time of its independence from Britain to 74% in 2011, but there is still a wide gap between men and women and between those who live in urban and rural areas.

In March of 2017, the United Nations estimated India's population as 1,337,756,334, and the national currency, the rupee, had the following exchange rate with the U.S. dollar: 1 dollar = 65.78 rupees. We will use this data in this activity.

a.   The World Bank estimates that 32.7% of Indian people live on less than $1.25 per day. Calculate the number of people living in India who try to survive each day with less than $1.25 to spend. Round your answer to the nearest whole.

b.   Suppose a man manages to save $1.25 of extra spending money per month. If a basic pair of shoes cost 700 rupees, then how many months will it take him to save enough money to buy the shoes? Round your answer to one decimal place.

c.   A woman who makes 2,500 rupees per month has the goal of saving enough money in one year to buy a television worth 21,000 rupees. How much money must she save each month in order to reach her goal? How much money does she have left each month to pay bills and live on? What is this amount in U.S. dollars? Round this amount to the nearest cent.

Monthly Savings (rupees):

Monthly (rupees) to live on: Monthly ($) to live on:

d.   India's two largest religions are Hinduism and Islam. If Hindus outnumber Muslims 5.7 to 1 and 80% of the country is Hindu, then what percent of India's population is Muslim? Round the percent to the nearest whole percent and populations to the nearest whole.

Percent of Muslims in Population:

Number of Hindus in Population:

Number of Muslims in Population:

e. For each year listed in the table, compute India's share of world population as a percent. Round each percent to two decimal places. Describe any trends you notice.

Year	Population of India	World Population	India's Share of World Population (%)
2017	1,342,512,706	7,515,284,153	
2016	1,326,801,576	7,432,663,275	
2010	1,230,984,504	6,929,725,043	
2000	1,053,481,072	6,126,622,121	
1990	870,601,776	5,309,667,699	
1980	697,229,745	4,439,632,465	
1970	553,943,226	3,682,487,691	
1960	449,661,874	3,018,343,828	

(*Source:* Worldometers)

f. For each pair of years listed in the table, use the data for Population of India given in the previous table to compute the amount of increase in India's population from the first year listed to the second year listed. Then calculate the associated percent of increase. Round each percent to two decimal places.

From Year	To Year	Amount of Increase in India's Population	Percent of Increase
2016	2017		
2010	2017		
2000	2017		
1990	2017		
1980	2017		
1970	2017		
1960	2017		

3.   Conceptual Exercise—Where's the Beef?

A supervisor at a meat packing plant needs to maintain a certain level of quality control for the product due to federal law. Suppose that the supervisor is in charge of marinated beef. In that department, workers place up to 500 pounds of meat into a tumbling machine and add a certain amount of marinade. To maintain quality, the final product cannot have more than a 20% increase in weight due to adding marinade.

a. An employee named John uses his memory of past experiences to estimate the amount of marinade added. John has just processed 200 pounds of beef, and the final product of marinated meat weighs 234 pounds. If a meat inspector makes a surprise visit, will the inspector give the company a violation? Show all work, and write your answer in complete sentences.

b. Suppose that 250 pounds of newly arrived beef (called the green product) is placed in a tumbling machine, and some marinade is added before packaging the beef (the final product) for shipment. What is the maximum possible weight of the marinated meat, if it is to stay within federal guidelines? Show all work and write your answer in complete sentences.

c. The meat inspector arrives at the plant and asks the supervisor to explain what method the plant employees use to calculate the 20% rule. This is the process that he describes:

1) Weigh the green product and record the value in pounds.
2) Take the weight in pounds and move the decimal point of this number one place to the left.
3) Double the number from Step 2 to arrive at the maximum increase in weight.
4) The final product is discarded if it weighs more than the green product plus the maximum increase just found in Step 3.

Does this procedure correctly meet the 20% rule required by federal law? Support your written answer by showing how the calculations work on 250 pounds of green product meat.

d. The meat inspector wants the supervisor to create a percent change formula that involves using the words *final product* and *green product*. An employee should be able to substitute the weight of the green product and final product into the formula, calculate the percent change, and then compare the value with 20%. Write this percent change formula below.

e. Test your percent change formula from part *d.* Use the formula in the following two situations.

- A plant employee weighs a fresh piece of meat at 230 pounds before mixing it with marinade. After the marinade is added, the weight of the beef is 272 pounds. Would the final product pass inspection under federal law?

- The same employee weighs a second portion of meat at 235 pounds before adding the marinade. After mixing it with marinade, the new weight is 284 pounds. Would the final product pass inspection under federal law?

4.   Group Activity—M&M's®

In the 1930s, many American stores did not stock chocolates during the summer because, without widespread air conditioning, the chocolate tended to melt and sales declined. In 1940, Forrest E. Mars, Sr., set out to develop a chocolate candy that could be sold year-round without the melting problem. So he formed a company in Newark, New Jersey, to make bite-sized melt-proof chocolate candies encased in a thin sugar shell. Thus M&M's Plain Chocolate Candies were born! The first M&M's were made for the U.S. military in 1941.

The first Plain M&M's colors were brown, green, orange, red, violet, and yellow. Violet was replaced by tan in 1950. When the safety of a particular type of red food coloring was publicly questioned in 1976, red was completely eliminated from the M&M's color mix to avoid alarming consumers, even though the coloring in question was never used in M&M's. After an 11-year hiatus, red returned to the color mix in 1987. In 1995, over 10 million Americans responded to a marketing campaign asking for help in choosing a new M&M's color. Given the choices of blue, pink, purple or no change, 54% of the respondents chose blue, and it replaced tan. At that time, the new color mix of Plain M&M's became blue (10%), brown (30%), green (10%), orange (10%), red (20%), and yellow (20%).

According to Mars, Incorporated, the current distribution of M&M's colors depends on which of two different U.S. factories produces them. The M&M's factory in Hackettstown, New Jersey, produces M&M's with the following official color distribution:

Blue	Brown	Green	Orange	Red	Yellow
25.0%	12.5%	12.5%	25.0%	12.5%	12.5%

a. Open a single-serving pouch of M&M's Plain Chocolate Candies. Count the blue, brown, green, orange, red, and yellow M&M's in your pouch and record the data in table below.

Blue	Brown	Green	Orange	Red	Yellow

b. What was the total number of M&M's in your pouch?

c. Find the percentage of each color in your pouch and record them in the table below. Round each percent to one decimal place.

% Blue	% Brown	% Green	% Orange	% Red	% Yellow

d. How do the percents you calculated in part *c* compare to the current official color distribution at the Hackettstown plant?

e. Combine your group's color counts with those of the other groups in your class. Use the following table to organize your data. Find candy totals for each color and the overall total number of M&M's tallied by all of the groups.

Group	Blue	Brown	Green	Orange	Red	Yellow	Total
1							
2							
3							
4							
5							
6							
7							
8							
9							
10							
TOTALS							

f. Find the percentage of each color for the combined data in part *e*, and report your results in the table below.

% Blue	% Brown	% Green	% Orange	% Red	% Yellow

g. How do these percentages for the combined data compare to the official color mix at the Hackettstown plant? Do the combined data give color percentages that are more similar or less similar to the official percentages than those you computed for a single pouch in part *c*? Explain.

1.   Extension Exercise—Winning BigBucks Lottery

Steve and Kate are wondering what 6 numbers to choose for Saturday's lottery game called BigBucks. This is a 6/49 lottery, meaning you play by choosing 6 different numbers between 1 and 49 inclusive and hope that your selection matches the winning combination (randomly picked by the State Lottery Commission).

Steve uses a number obtained by calling 1-900-PSYCHIC to try and win the $2,674,763 jackpot. Kate decides on the following 6-number combination: 1, 2, 3, 4, 5, 6.

Her logic is that since any 6-number combination is equally likely to occur, the above pick has as good a chance of winning as any other combination.

a. Let's start by exploring the six-number combination you must select to play a 6/49 lottery. How many possible numbers are available when choosing the first number?

b. Once the first number is chosen, how many numbers do you have available when choosing a second number (different from the first number)?

c. How many ways are there to choose the first two numbers? Write your answer as a product of two numbers, and then compute the result.

d. After the first two numbers are chosen, how many numbers are available when choosing a third number (different from the first and second numbers)?

e. How many ways are there to choose the first three numbers? Write your answer as a product of three numbers, and then compute the result.

f. Knowing that there are $49 \cdot 48 \cdot 47 = 110,544$ ways to choose the first three numbers, how many ways can you choose the entire winning six-number combination? Write your answer as the product of six numbers, and then compute the result.

In the last six questions, we have treated the order of appearance of each number as being important (different). For example, we have counted combinations such as 48   7   16 as being different from 7   16   48.

In fact, when playing the lottery, the order in which the numbers appear is not important, so we have over-estimated the number of ways to choose the entire six-number combination necessary to win BigBucks lottery. Let's try to figure out what value must be divided out, so that we are counting only the different combinations without regard to order.

g. Answer the following questions to find how many ordered arrangements (or permutations) of 6 numbers exist.

- Given six distinct numbers, how many ways can you choose the first number? _____

- Having chosen the first number, how many ways can you choose the second number?

- Having chosen the first and second numbers, how many ways can you choose the third number?

- Having chosen the first, second, and third numbers, how many ways can you choose the fourth number?

- Having chosen the first, second, third, and fourth numbers, how many ways can you choose the fifth number?

- Having chosen the first, second, third, fourth and fifth numbers, how many ways can you choose the sixth number?

- Altogether, how many ways can you order six numbers?

h. Find the quotient of the product from part *f* divided by the product from part *g*. The result is all the possible six number combinations that have an equally likely chance to occur.

$$\text{Number of possible outcomes} = \frac{product\ from\ part\ f}{product\ from\ part\ g} =$$

i. If a favorable outcome is selecting the winning number with your ticket, then how many favorable outcomes exist?

Number of favorable outcomes from your ticket:

j. Use the probability ratio to find the chance (probability) of your ticket winning the BigBucks lottery game.

$$\textit{Probability of winning} = \frac{Number\ of\ favorable\ outcomes}{Number\ of\ possible\ outcomes} =$$

2.   Exploration Activity—What Is Average?

Mr. Smith is ready to give back graded midterm exams to his class of 32 students. A student asks him, "How did the class do?" Mr. Smith displays the following chart on the overhead projector.

56	72	74	67	74	92	81	76
82	90	75	98	61	88	65	94
63	58	82	50	77	83	78	86
69	82	61	75	85	78	61	72

a. Mr. Smith asks you to complete the table given below.

Class Intervals (Exam Scores)	Class Frequency (Number of Students within given grade range)
50–59	
60–69	
70–79	
80–89	
90–99	
TOTAL Number of Students	

b. Next Mr. Smith asks you to construct a histogram from the table in part *b*. Choose an appropriate scale for the vertical and horizontal axes.

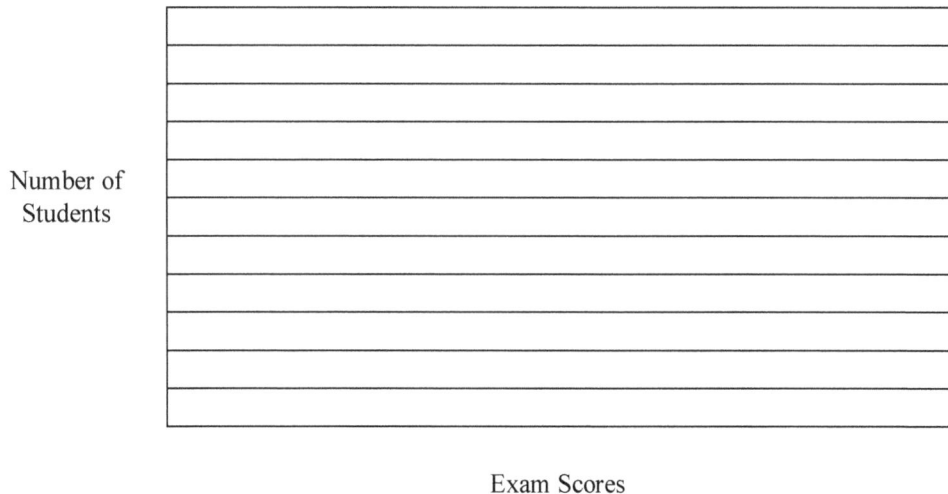

Number of Students

Exam Scores

c. Suppose that Mr. Smith uses the following system to convert numerical scores to letter grades:

      A: 90-100   *B*: 80-89   *C*: 70-79   *D*: 60-69   *F*: 0-59

Answer the questions below and round each to the nearest whole.

- What percent of the students in the class received an *A*?

- What percent of the students in the class received a *B*?

- What percent of the students in the class received a *C*?

- What percent of the class passed (60-100) the midterm?

- What percent of the students failed (0-59) the midterm?

d. Use the information from parts *a* and *c* to create a pie graph (circle graph) of the percentages of A's, B's, C's, D's, and F's in the class.

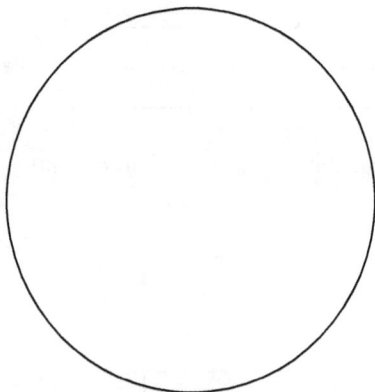

e. For the given set of exam scores,

- find the mean score of the class. Round to the nearest whole.

- find the median score of the class.

- find the mode score of the class (there may be more than one).

f. Which measures of central tendency best describe the exam scores of the class? Explain using complete sentences.

3.   Conceptual Exercise—Lightning Strikes

Suppose we want to compare the probability of winning a 6/49 lottery with the probability, over a one-year time period, of a randomly selected American being struck by lightning. According to Worldometers, the resident population of the United States is approximately 325,000,000. According to the National Oceanic Atmospheric Administration, an average of 310 people are struck by lightning in the United States each year.

a. What is the probability that a randomly selected American resident will be struck by lightning this year? State the probability as a fraction, a decimal rounded to a single non-zero digit, and a percent. Use complete sentences to explain your answers.

Fraction:

Decimal:

Percent:

b. Do you think that every resident of the United States has the exact same probability of being hit by lightning? Use complete sentences to explain your answer.

c. If you round the chances of winning the lottery to 1 in 14 million, then how many more times likely are you to be hit by lightning than to win the lottery? Show all work and explain your answer using complete sentences.

4. Group Activity—Laundry Blues

Tom woke up on a rainy Monday morning with nothing to wear for class. He forgot to do his laundry yesterday. It turns out that his roommate John actually did some of his laundry for him. There are 4 clean shirts and 3 clean pairs of pants hanging in the closet. Tom thinks that he doesn't have enough choices to mix and match into a suitable outfit to wear, but John says that there are 12 possible outfits to choose from, and at least one of these outfits must match!

a. Let's define the 4 clean shirts as S#1, S#2, S#3, and S#4, and the 3 pairs of clean pants as P#1, P#2, and P#3. Use a tree diagram to list the 12 possible outfits that Tom can choose from.

b. Suppose there are 8 possible combinations that Tom would consider suitable. If he walks into the dark closet and randomly chooses one pair of pants and one shirt without being able to distinguish colors or styles, then what is the probability that Tom has picked an outfit that he considers to be suitable? Show all work, and state your answer as a fraction, a decimal rounded to the hundredths place, and a percent.

    Fraction: Decimal: Percent:

c. What is the probability that in the dark closet Tom has randomly picked out an outfit that is not suitable? Show all work, and state your answer as a fraction, a decimal rounded to the hundredths place, and a percent.

    Fraction: Decimal: Percent:

d. Assume that S#1 is Tom's favorite shirt. What is the probability of randomly picking this shirt in the dark closet? Show all work, and state your answer as a fraction, a decimal rounded to the hundredths place, and a percent.

    Fraction: Decimal: Percent:

e. Suppose there is one shirt with two buttons missing and 1 pair of pants with a broken zipper, what is the probability that Tom has picked:

- both these items?

    Fraction: Decimal (to nearest hundredths place):   Percent:

- one of them?

    Fraction: Decimal (to nearest hundredths place):   Percent:

f. Tom finally chooses an outfit. While getting dressed, he listens to the radio and hears that there is a 75% chance of rain. Does this mean that rain will fall from the sky 75% of the time that day? Support your answer using complete sentences.

1. Extension Exercise—Survival of the Hatchlings

There are about 29,000–40,000 adult female Leatherback sea turtles in the world. On average, a female turtle lays eggs every 2–3 years. She nests 4 or 5 times in one season, laying about 90 eggs per nest. Approximately 40 percent of the eggs hatch, and 1 in 2500 hatchlings survive to adulthood.

a. Since female Leatherbacks lay eggs on average every 2 to 3 years, we could roughly guess that in one year about one-half to one-third of the females are nesting. Calculate an estimate of the least number of females that nest in one year, assuming that $\frac{1}{3}$ of the lowest estimate, 29,000 adult females, nest in that year. Round your answer to the nearest whole number. Show your work and answers below.

Least number of females that nest in a year: _____

b. Calculate an estimate of the greatest number of females that nest in a year, assuming that $\frac{1}{2}$ of the highest estimate, 40,000 adult females, nest each year. Show your work and answers below.

Greatest number of females nesting in a year: _____

c. Use the information from parts *a* and *b* to show the probable range of Leatherback female turtles that nest each year.

From _____ to _____ Leatherback female turtles nest each year.

d. Determine a range of how many eggs are expected to hatch from one female Leatherback turtle in one season (if this is her nesting year). Show your work.

Hint: number of nests per year for one turtle × number of eggs per nest × 40% = number of turtles hatched per one female Leatherback turtle

From _____ to _____ eggs are expected to hatch from one female turtle in one season.

e. Review the information in parts *c* and *d*, and record the range of expected total number of hatchlings in any year.

Hint: total number of hatchlings = number of female turtles that nest × number of eggs that are expected to hatch per turtle.

From _____ to _____ Leatherback turtle eggs are expected to hatch each year.

f. If the survival rate of hatchlings is 1 : 2500, then what is the range of each year's hatchlings that are expected to survive to adulthood? Set up and solve proportions, one for the low end of the range and one for the high end of the range, to answer the question. Round each answer to the nearest whole.

From _____ to _____ Leatherback hatchlings are expected to survive to adulthood.

g. Mature Leatherbacks measure anywhere from 1.2 to 1.9 meters long and weigh from 200 to 500 kilograms. The heaviest Leatherback ever recorded weighed 916 kilograms. Convert these metric measurements and weights to the approximate number of units in the U.S. system. Rewrite the first two sentences of this paragraph using U.S. system equivalents. Round your U.S. units to the nearest tenth for length and to the nearest whole number for weight. Show the unit fractions and set-ups in your conversions.

Hint: 1 meter ≈ 3.28 feet (≈ means "approximately equals"); 1 kilogram ≈ 2.20 pounds

1.2 meters ≈ _____

1.9 meters ≈ _____

200 kilograms ≈ _____

500 kilograms ≈ _____

916 kilograms ≈ _____

2.   Exploration Activity—Fence Me In

Suppose you just bought a house that sits on a rectangular lot with an area of 7200 square feet. The house is in good shape, but you want to enclose the borders with a fence. You know that the lot is 7200 square feet, and yet you still don't know the actual length or width of the lot.

a.   Use the given data below to complete the table and list some possible dimensions for your 7200 square foot lot. Draw a picture of the situation to help you visualize the problem.

$$Area = Length \cdot Width$$
$$Perimeter = 2 \cdot (Length + Width)$$

Table with some given lengths:

Area	Length	Width	Perimeter
7200	40		
7200	60		
7200	72		
7200	80		
7200	90		
7200	100		
7200	120		
7200	180		

b.   In the above table, what happens to the width as the length increases? Answer using complete sentences, and explain any patterns you observe.

c.   In the table above, what happens to the perimeter as the length increases? Answer using complete sentences, and explain any patterns you observe.

d. Suppose you want to enclose the entire perimeter (border) of your property with 6 foot high fencing that cost $5.00 per foot of length purchased. The total length of fence you need depends on the perimeter. If you could choose any length/width pair from the previous table, what dimensions would minimize the cost of fencing in your property? Justify your answer using complete sentences.

e. Do you think there are dimensions not listed in the table that would decrease the cost of fencing even more than your answer to part *d*? If yes, approximate the length and width. Explain your answer using complete sentences.

f. Let's investigate by looking at another table that starts with a length of 80 ft and goes up in 1-foot increments to a length of 90 feet. Fill in the columns for width, perimeter, and fencing cost using the equations below. Round results to the nearest hundredth.

$$Width = \frac{7200}{Length} \qquad Perimeter = 2 \cdot (Length + Width) \qquad Fencing\ Cost = \$5 \cdot (6 \cdot Perimeter)$$

Area	Length	Width	Perimeter	Calculations	Fencing Cost
7200	80				
7200	81				
7200	82				
7200	83				
7200	84				
7200	85				
7200	86				
7200	87				
7200	88				
7200	89				
7200	90				

g. Which perimeter gives the lowest fencing cost? Which dimensions would result in the lowest fencing cost? What is the lowest fencing cost?

3.   Conceptual Exercise—Converting Doses

A pharmacy's stock of amoxicillin includes 20 bottles of amoxicillin, containing 100 capsules in each bottle, plus a partially full bottle containing 25 capsules. Each capsule holds 250 milligrams of amoxicillin.

    Information: 1 kilogram = 1000 grams   and   1 gram = 1000 milligrams

a. How many total milligrams of amoxicillin does the pharmacy have in stock? Convert your result to both grams and kilograms. Round results to the nearest hundredths place.

    amoxicillin in stock: _____ mg = _____ g = _____ kg

b. How many milligrams are equivalent to 1 kilogram? *Hint:* Use the information given at the beginning of this activity to set up the necessary unit fractions to make the unit conversion.

        1 kg = _____ mg

c. Doctor Alvarez prescribed 750 milligrams of amoxicillin per day for seven days. How many total milligrams did the doctor prescribe? Convert the total to grams.

    amoxicillin prescribed: _____ mg = _____ g

d. Since there are 250 mg of amoxicillin in each capsule, how many capsules should the pharmacist put in the bottle to fill the prescription described in part *c*?

                        _____ capsules

e. If the prescription from part *c* is followed, how many capsules should be taken in one day?

Take _____ capsules per day

f. The pharmacist must make up a label to tell the patient how many capsules to take and when. If the capsules are to be taken in equally spaced intervals over 24 hours, then what instructions should be written on the label? *Hint:* Tell the patient how many capsules to take at which points during the day.

g. How much amoxicillin is left in stock after the prescription (from part *c*) is filled? Express your answer in milligrams as well as in grams.

amoxicillin in stock: _____ mg = _____ g

4. Group Activity—Composting

To keep your garden soil rich with organic matter, you decide to keep your yard waste in a compost pile. Building a bin using extra wood and wire will promote faster decomposition. A good compost bin is shaped like a cube and has a volume ranging from 27 to 125 cubic feet ($ft.^3$). Work with your group members to answer the following questions.

Materials needed: paper, scissors, ruler

a. A die that you might toss in a board game is an example of a cube. How many numbered faces (sides) does it have?

b. Each face on a cube is an identical square. Represent the compost bin by drawing a wire frame cube that shows all the edges (lines) that would make up the 6 faces.

c. What is the range of length measurements for each identical face (side), if we are to stay within the recommended limits for the volume of the compost bin? *Hint:* $ft \times ft \times ft = ft^3$.

d. Let's make two scaled-down models of the smallest and largest possible bins (with no lid). Use a piece of paper along with scissors and a metric ruler to construct a scaled down model of our cubic bin, with a measure of 1 inch being equivalent to 1 foot of the real bin. To model the smallest recommended bin, use the dimensions, $3\,in \times 3\,in \times 3\,in$, giving a volume of 27 cubic inches (in^3). Follow the steps below.

1.) Draw a square with each of the 4 sides having a length of 9 in.

2.) From each of the 4 corner points (vertices), measure and mark 3 inches in the horizontal direction and 3 inches in the vertical direction.

3.) Use your ruler to draw a 3 inch by 3 inch square in each of the 4 corners.

4.) Use your scissors to cut out each 3 inch by 3 inch square in each of the 4 corners.

5.) Fold up all of the sides to form an open box that will model our compost bin.

Next, model the largest possible bin using the dimensions, 5 inches by 5 inches by 5 inches giving a volume of 125 cubic inches.

e. Which of the following two choices would be the most economical for composting? Answer using complete sentences and support your answer with mathematical reasoning.

- 1 compost bin with a volume of 125 cubic feet

OR

- 4 compost bins each with a volume of 27 cubic feet

f. Suppose you only have enough material to make a compost bin measuring $3\frac{1}{2}$ ft $\times 3\frac{1}{2}$ ft $\times 3\frac{1}{2}$ ft. How many cubic feet of lawn waste could it hold?

g. Suppose you can carry $\frac{1}{2}$ cubic foot of lawn waste in each full shovel. If you end up with 80 full shovels of lawn waste, will the bin that you constructed in part *f* overflow? Explain.

h. What are the dimensions of a cubic compost bin that has a volume of 64 cubic feet? Explain how you found these dimensions.

1. Extension Exercise—How Hot Is It?

In the United States, temperature is generally reported in degrees Fahrenheit. Most of the world, however, measures temperature in degrees Celsius. During the Olympics, for example, the temperatures are reported in degrees Celsius. We may have to convert from degrees Celsius to degrees Fahrenheit (or the reverse) in order to appreciate how hot or cold it is outside.

Formula for converting from degrees <u>Celsius *C* to Fahrenheit *F*</u>: $F = 32 + \dfrac{9}{5}C$

a. Solve the formula $F = 32 + \dfrac{9}{5}C$ for *C* to obtain a formula for converting from degrees Fahrenheit (F) to degrees Celsius (C).

b. Fill in the table below by converting between degrees Celsius and degrees Fahrenheit. Round your answers to the nearest tenth of a degree.

Celsius Temperature	Fahrenheit Temperature
−15°	
9°	
32°	
63°	
115°	
	−31°
	0°
	15°
	50°
	100°
	212°

The coldest temperatures (in degrees Fahrenheit) ever recorded in the United States are shown in the table below.

Year	State	Degrees Below 0°F	Coldest Temperature (F) Written as an Integer	Celsius Temperature Equivalent
1933	Wyoming	66		
1936	North Dakota	60		
1943	Idaho	60		
1954	Montana	70		
1971	Alaska	80		
1985	Colorado	61		
1985	Utah	69		
1996	Minnesota	60		

(*Source:* National Climatic Data Center)

Answer the following questions based on the table above.

c. Fill in the fourth column of the table above by using an integer to express the coldest temperature for each state.

d. Fill in the fifth column of the table above by using the formula from part *a* to convert each Fahrenheit temperature in the fourth column to Celsius.

e. What was the lowest temperature ever recorded in the United States?

f. In what year and state did the lowest recorded temperature occur?

2. Exploration Activity—Fruit Salad

You've started a catering business and have been asked to quote a price per person for catering fruit salad at an afternoon reception for 20 people. The recipe you will use is given below.

Fruit Salad
  2 pounds green grapes, halved
  3 pounds watermelon, cubed
  1 pint fresh blueberries
  $1\frac{1}{4}$ pounds nectarines or mangos, sliced
  1 6-ounce can frozen pineapple juice concentrate
Combine ingredients and serve chilled. Serves 10 guests.

The wholesale prices for the ingredients are as shown below.
  green grapes   84 cents per pound
  watermelon   21 cents per pound
  blueberries   92 cents per pint
  mangos   50 cents per pound
  pineapple juice concentrate 43 cents per 6 oz. can

a. Enter the quantity and unit price of each ingredient into the table below to compute the price of each item and the total cost for all of the ingredients. (Recall that you are serving 20 people.)

Ingredient	Quantity	Unit Price	Price
Green grapes			
Watermelon			
Blueberries			
Mangos			
Pineapple juice concentrate			
			Total Cost:

b. *Cost per serving* is the amount you pay, per person, for your raw materials. Using the information from the table above, write an equation that relates the cost per serving, *C*, to the total cost for all of the ingredients. (Remember that the salad will serve 20 people.)

c. Estimate the time it will take you to buy the ingredients and to make the fruit salad once you have the ingredients. Record your estimates in hours or fractions of hours.

Estimated time for trip to grocery store (roundtrip plus shopping time): _____

Estimated time to make the fruit salad: _____

d. The salad must be delivered, and you estimate it will take about an hour to deliver the salad and set up the table. Later you will return to pick up the salad bowl and clean the table, which you estimate will take about a half-hour. Compute the total time you estimate it will take to buy ingredients, make the salad, deliver the salad, and pick up the salad bowl.

Total time spent: _____

e. Since you are trying to make catering your sole business, you determine that you need to make about $20 an hour for your time on each catering event. Use the result of part *d* to determine how much you would have to earn in profit (over the cost of the salad) to make $20 an hour for your time.

f. Letting *s* be price charged per serving of fruit salad, *C* be the total cost of the fruit salad ingredients, *P* be the total profit you hope to earn, and *n* be the number of people to be served, write a formula for the price that should be charged per serving for the fruit salad. Then use your formula and your results from parts *a* and *e* to determine how much you should charge per serving for the salad if you have been asked to serve 20 people.

3. Conceptual Exercise—Freezing/Melting Point and Boiling Point Temperatures

a. The freezing point of water is 32°F. This means that water is in its solid state (ice) for Fahrenheit temperatures that are _____ (less than / greater than) 32°F.

b. Let the variable F represent Fahrenheit temperatures. Write an inequality that describes the range of all Fahrenheit temperatures for which water is in its solid state. Then write the solutions in interval notation.

c. Recall that the formula $F = 32 + \dfrac{9}{5}C$ describes the relationship between Fahrenheit temperatures F and Celsius temperatures C. Substitute $32 + \dfrac{9}{5}C$ for F in your inequality from part *b*, and solve it for C. Interpret the meaning of the resulting inequality. Then write the solutions in interval notation.

d. The boiling point of water is 212°F. This means that water is in its gaseous state for Fahrenheit temperatures that are _____ (less than / greater than) 212°F.

e. Write an inequality that describes the range of all Fahrenheit temperatures F for which water is in its gaseous state. Then write the solutions in interval notation.

f. Use the formula $F = 32 + \dfrac{9}{5}C$ to make a substitution in your inequality from part *e*, and solve it to find an inequality that describes the range of Celsius temperatures C for which water is in its gaseous state. Then write the solutions in interval notation.

g. The melting point of gold is 1948°F. Use an inequality to find the range of Celsius temperatures for which gold is solid.

h. The boiling point of sodium is 1621°F. Use an inequality to find the range of Celsius temperatures for which sodium is in its gaseous state.

4. Group Activity—Zoya's Investments

Zoya would like to invest some money so that she earns <u>at least</u> $850 in simple interest per year. Let's look at two different scenarios for this situation.

SCENARIO A

In this scenario, Zoya has received a $40,000 bonus. She would like to invest all of it to earn at least $850 in simple interest per year. There are many different investment options available to her, each with different interest rates. What is the lowest interest rate at which Zoya could invest the full amount of her bonus to achieve her earnings goal of at least $850 per year?

<u>Step 1: UNDERSTAND the Problem.</u>

a. Read and reread the problem. What formula could be used to help solve this problem? Define the variables used in it.

b. Before continuing, let's become familiar with the formula you chose. Use the formula to complete the following table for how much simple interest is earned on $40,000 after 1 year when invested at the various interest rates shown.

Interest rate	Simple Interest earned after 1 year on $40,000
0.10%	
0.75%	
1.0%	
1.5%	
2.0%	
2.75%	
3.5%	

<u>Step 2: TRANSLATE the problem.</u>

c. Write an inequality based on the formula you chose in part *a* to aid in solving the word problem.

Step 3: SOLVE.

d. Solve the inequality from part *c*. Write the solution set in interval notation.

Step 4: INTERPRET the results.

e. Check the proposed solution in the stated problem and state your conclusion. Is your result in line with the table you completed in part *b*? Explain.

SCENARIO B

In this scenario, Zoya would like to invest an amount of money so that she can earn at least $850 in simple interest per year. She has selected an investment that pays 1.8% simple interest. What is the minimum amount she should invest to achieve her earnings goal of at least $850 per year?

Step 1: UNDERSTAND the Problem.

f. Read and reread the problem. What formula will you use to help solve this problem?

g. Before continuing, let's become familiar with the formula you chose. Use the formula to complete the following table for how much simple interest is earned on various principal amounts after 1 year when invested at 1.8% simple interest.

Principal Amount	Simple Interest earned after 1 year at 1.8%
$10,000	
$20,000	
$30,000	
$40,000	
$50,000	
$60,000	

Step 2: TRANSLATE the problem.

h. Write an inequality based on the formula you chose in part *f* to aid in solving the word problem.

Step 3: SOLVE.

i. Solve the inequality from part *h*. Write the solution set in interval notation.

Step 4: INTERPRET the results.

j. Check the proposed solution in the stated problem and state your conclusion. Is your result in line with the table you completed in part *g*? Explain.

1. Extension Exercise—Sibling Running Rivalry

Nia and her younger brother Amari have challenged one another to two running races. To help Amari overcome the disadvantage of his smaller stature, Nia agrees to allow Amari to begin each race from a point that is 100 meters ahead of her starting line. At the starter's signal, both Nia and Amari will start running at the same time toward a fixed finished line, and whoever arrives there first wins the race.

The following two equations describe each runner's position d (in meters past Nia's starting line) as a function of time t (in seconds after the starter's signal).

$$\text{Nia's position: } d = 10t$$

$$\text{Amari's position: } d = 7t + 100$$

Use these two equations to help answer the following questions.

a. In the siblings' first race, the finish line is 200 meters past Nia's starting line. Who will win the first race? What is the winning time? Round to two decimal places. (*Hint:* Set each position equation equal to 200 meters and solve for t. Low time wins.)

b. Using the winning time for the first race from part *a*, calculate how far (in meters) the winner will be ahead of the loser when the winner crosses the finish line.

c. In the siblings' second race, the finish line is 400 meters past Nia's starting line. Who will win the second race? What is the winning time? Round to two decimal places.

d. Using the winning time for the second race from part *c*, calculate how far (in meters) the winner will be ahead of the loser when the winner crosses the finish line.

e. After how many seconds will Nia overtake Amari in the second race? Round to two decimal places.

f. Assuming that Nia and Amari continue running according to the given equations, explain why Nia will always catch up to Amari if they run long enough. What part of the linear equations immediately tells you this fact?

2. Exploration Activity—Income and Tax Liability

Consider the following 2019 federal tax rate schedule, based on data from the Internal Revenue Service, for people whose filing status is single. Note: Any reference to *income* means *taxable income*.

If your income is		Your tax liability (tax owed) is	Of any income over-
Over-	But not over-		
$0	$9,700	-------------- 10%	$0
9,700	39,475	$970 + 12%	9,700
39,475	84,200	4,543 + 22%	39,475
84,200	160,725	14,382.50 + 24%	84,200
160,725	204,100	32,748.50 + 32%	160,725
204,100	510,300	46,628.50 + 35%	204,100
510,300	---------	153,798.50 + 37%	510,300

Your goal is to develop a set of equations that will tell us how much federal tax must be paid at different income levels. Suppose x represents a person's income and y is the tax liability (tax that must be paid to the federal government on that income). The equations you will construct should express tax liability as a function of income.

a. The first row of numbers in the tax rate schedule means that if your income is between $0 and $9,700, then your tax liability is 10% of your income. Fill in the table below, which calculates the tax liability for a range of incomes within this first bracket.

Income, x	Tax Liability, y
$0	
1,500	
3,000	
4,500	
6,000	
7,500	
9,000	

b. Explain the pattern in the table by completing the following sentence.

As the income increases by $1500, the tax liability _____

_____ .

c. Select two ordered pairs, (x, y), from the data table in part *a*, and use them to calculate the slope. Express your slope as a decimal.

d. Explain the connection between the slope and the tax rate.

e. Plot all of the data points on the graph below. Connect the data points to form a straight line.

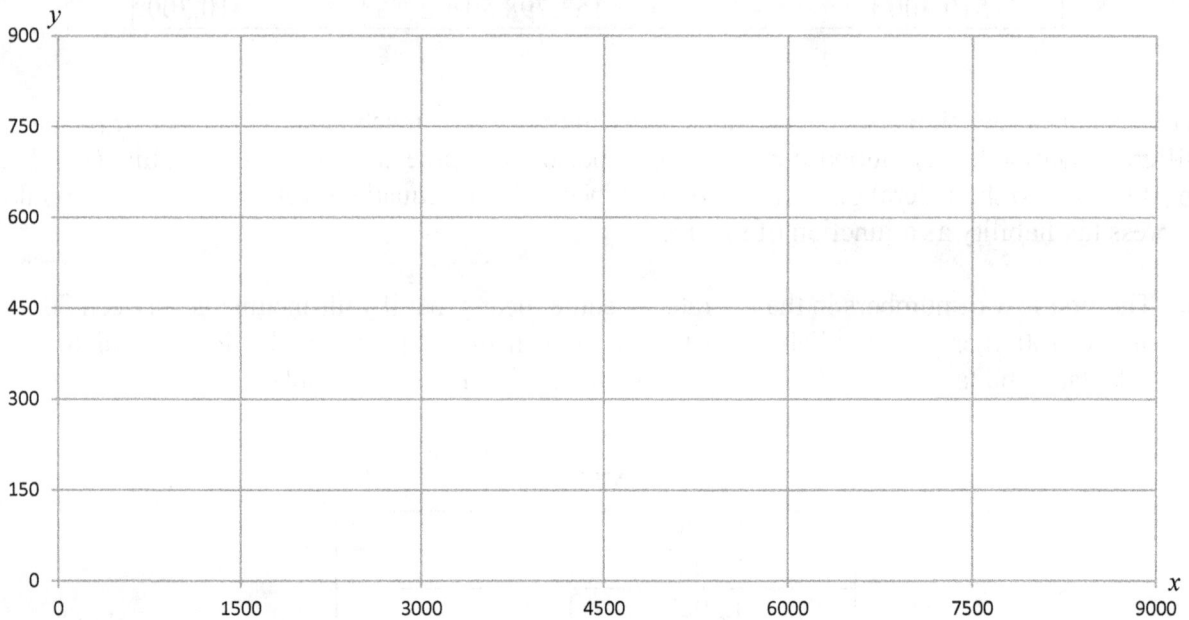

f. What is the vertical intercept point or y-intercept point? (_____ , _____)

g. A linear equation in two variables can be written in the form $y = mx + b$, where m is the slope and b is the y-intercept. What is the equation that represents tax liability y in terms of income x for the 10% tax bracket?

3. Conceptual Exercise—Road Grades

Recall that the **grade** of a road is its slope m written as a percent. This means that a road with a 3% grade rises (or falls) 3 feet for every horizontal 100 feet, because

$$3\% = 0.03 = \frac{3}{100} \quad \text{and} \quad \frac{\text{vertical change}}{\text{horizontal change}} = \frac{\text{rise}}{\text{run}} = m$$

According to the American Association of State Highway and Transportation Officials, interstate highways in the United States have a maximum allowable grade of 6%, although a 7% grade may be allowed in mountainous areas where the speed limit does not exceed 60 mph. However, these limits on road grade apply only to interstate highways. State highways and local roads (usually having lower speed limits than controlled-access interstates) often utilize higher road grades. For instance, the grade of Arizona Highway 77 at El Capitan Pass is 8%, and a portion of Wyoming Highway 22 at Teton Pass has a grade of 10%.

a. Suppose you start driving at sea level (an elevation of 0 feet) at the bottom of a hill with an 8% upward grade. If it were possible to drive 50 miles (measured horizontally) on this hill with an 8% grade, what would your elevation above sea level be after 50 miles? Give your answer in miles and then in feet. (*Hint:* 1 mile = 5280 feet, so 50 miles = [5280×50] feet.)

b. Outside of Boulder, Utah, a sign on Utah State Route 12 warns truck drivers to use low gears because there is a downward grade of 14% for the next 4 miles. If this 4-mile stretch were measured horizontally (rather than at an angle), what is the drop in elevation (in feet) over that 4-mile stretch of road? Show how you arrived at your answer and draw a diagram to illustrate the grade of the road. (*Hint:* Be sure to convert 4 miles to feet as shown in the hint for part *a*.)

c. Denver is called the "Mile High City" because its elevation is approximately 1 mile above sea level. Suppose you are at sea level and Denver is 1000 miles away (measured horizontally). If you are traveling along a highway with a steady incline, what would be the grade of the highway?

　　　　　　　　　　　　　　　　　　　　　　79

d. Pikes Peak is about 65 miles (measured horizontally) from Denver if you could get there on a straight road with no curves. Pikes Peak has an elevation of about 14,110 feet above sea level. If a road could be built from Denver to Pikes Peak with a steady incline and no curves, what would the grade of the road be? Round the grade percent to 1 decimal place.
(*Hint:* Convert miles to feet. Use a calculator as needed.)

e. The summit on US Highway 33 near Judy Gap, West Virginia, is at an elevation of 3595 feet above sea level. The town of Franklin, West Virginia, has an elevation of about 1720 feet above sea level. The route that connects them is about 9 miles (measured horizontally). If the incline were the same everywhere on the road, what would be the approximate slope of the road? Also report the road's slope as a grade, rounded to the nearest tenth.
(*Hint:* Use a calculator as needed.)

f. If a highway has a slope of $\dfrac{1}{7}$, would it satisfy the grade requirements for a U.S. interstate highway? What about a slope of $\dfrac{2}{37}$? Explain your reasoning.

g. Why do you think it is helpful to drivers for road grades to be posted in mountainous terrain?

4. Group Activity—Analyzing the Federal Income Tax Rate Schedule

Use the 2019 federal tax rate schedule for single taxpayers shown below to complete the questions in this activity.

If your income is		Your tax liability (tax owed) is	Of any income over-
Over-	But not over-		
$0	$9,700	-------------- 10%	$0
9,700	39,475	$970 + 12%	9,700
39,475	84,200	4,543 + 22%	39,475
84,200	160,725	14,382.50 + 24%	84,200
160,725	204,100	32,748.50 + 32%	160,725
204,100	510,300	46,628.50 + 35%	204,100
510,300	---------	153,798.50 + 37%	510,300

The second row of numbers from the table can be translated as follows: If your income is over $9,700 but not over $39,475, your tax liability is $970 plus 12% of any income over $9,700.

a. Work with your group members to fill in the table below, which calculates the tax liability for several incomes within this second bracket.

Income, x	Amount over $9,700	Tax Calculation	Tax Liability, y
$10,000			$1006
14,000			
18,000			
22,000			
26,000			
30,000			
34,000			
38,000			

b. If x is the income, then write an expression for calculating "12% of income over $9,700" in terms of x.

c. Work with your group members and use the information from part *b* to find the linear equation that represents tax liability *y* in terms of income *x* for the 12% tax bracket. Simplify your equation so that it is in the form $y = mx + b$.

d. Each group member should choose two ordered pairs, (x, y), from the data table in part *a* and use them to calculate the slope. Try to have each group member select a different set of ordered pairs and compare results with the group.

e. What is the connection between your group's slope calculations in part *d* and the coefficient *m* found in part *c*? What does the slope mean in terms of the problem situation?

f. Your group is given the task of finding equations (formulas) for each of the remaining tax brackets. Decide as a group the best way to accomplish this task, and then proceed.

22% Bracket:

24% Bracket:

32% Bracket:

35% Bracket:

37% Bracket:

g. Substitute each income listed below into the appropriate formula from part *f.* Then calculate the tax liability for a single person with that income.

- $50,000

- $100,000

- $200,000

- $400,000

- $600,000

h. How does the U.S. tax system affect different incomes? Explain the advantages (if any) and disadvantages (if any) of moving up to a higher income tax bracket.

1. Extension Exercise—Signing the Best Contract

Suppose that Megan wrote a novel during her last year in school, and two publishers extend her offers to sign the manuscript for publication. Company A offers her a $10,000 sign-on bonus and 15% of the book's total sales. Company B offers her a $5000 sign-on bonus and 18% of the book's total sales. Assuming both companies would spend about the same amount on advertising and promoting the book, which publisher should she sign with? Complete this extension exercise to make a recommendation.

a. Complete the following table showing Megan's potential income as a function of her book's total sales as offered by each publishing company.

Total Sales ($)	Income from Company A ($)	Income from Company B ($)
$0		
25,000		
50,000		
75,000		
100,000		
125,000		
150,000		
175,000		
200,000		
225,000		
250,000		

b. Look at your table from part *a*. As the total sales of Megan's book increase, how do the incomes from Company A and Company B compare?

c. Plot the income from Company A as a function of total sales, connect the points with a straight line, and label it A. On the same graph, plot the income from Company B as a function of total sales, connect the points with a straight line, and label it B.

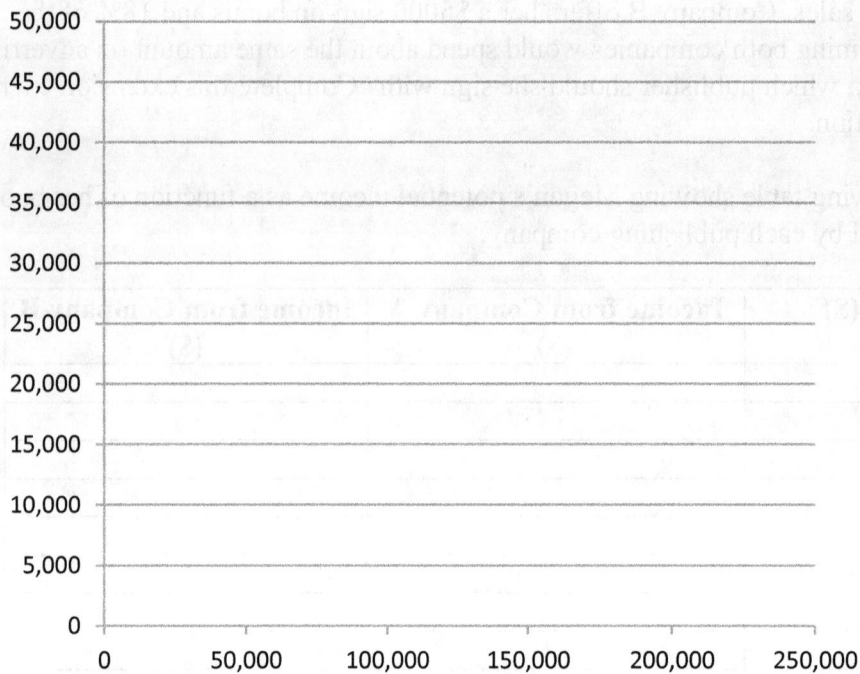

d. Describe the situation, including any trends you see, in the graph from part *c*.

e. Let *x* represent the total sales of her book (in dollars), and let *y* represent the income (in dollars) that Megan would receive. Using these variables, write an equation that relates income *y* to total sales *x* for Company A. Write another equation that relates income *y* to total sales *x* for Company B. Use these equations to form a system of equations for the two companies, and then solve it using the substitution method.

　　　Company A:　$y =$ _____

　　　Company B:　$y =$ _____

　　　System of equations:

　　　Solve your system using the substitution method:

f. Explain what the solution to the system means in terms of the problem situation.

　　　　　　　　　　　　　　　　　　　　　　　　87

g. For each equation in the system of equations formed in part *e*, explain the meaning of the slope and *y*-intercept in terms of the problem situation.

h. Based on all of the information, how would you advise Megan on selecting a publisher? Be sure to support your advice with mathematical reasoning.

2. Exploration Activity—The Price Is Right

Let us consider how the unit price p (in dollars) of some product can affect the quantity of that product that consumers will want to purchase (demand), as well as the quantity of that product that producers are willing to make available to purchase (supply). The supply function, s, is the quantity of a product that a producer will supply to the market at different prices p. The demand function, d, is the quantity of the product that consumers will demand at different prices p. Suppose a certain product has the market demand and supply functions given below.

$$d = 2000 - 10p$$
$$s = 5p - 500$$

a. Fill in the quantities supplied and demanded at the given prices in the table below. Explain why the prices from $0 to $80 are not valid prices for the supply function.

Price Per Unit (in dollars)	Quantity Demanded	Quantity Supplied
0		
20		
40		
60		
80		
100		
120		
140		
160		
180		
200		

b. Explain what a price of p = \$100 means in terms of supply and demand for this product.

Explain what a price of p = \$200 means in terms of supply and demand for this product.

c. Sketch a graph of the supply and demand functions on the same axes.

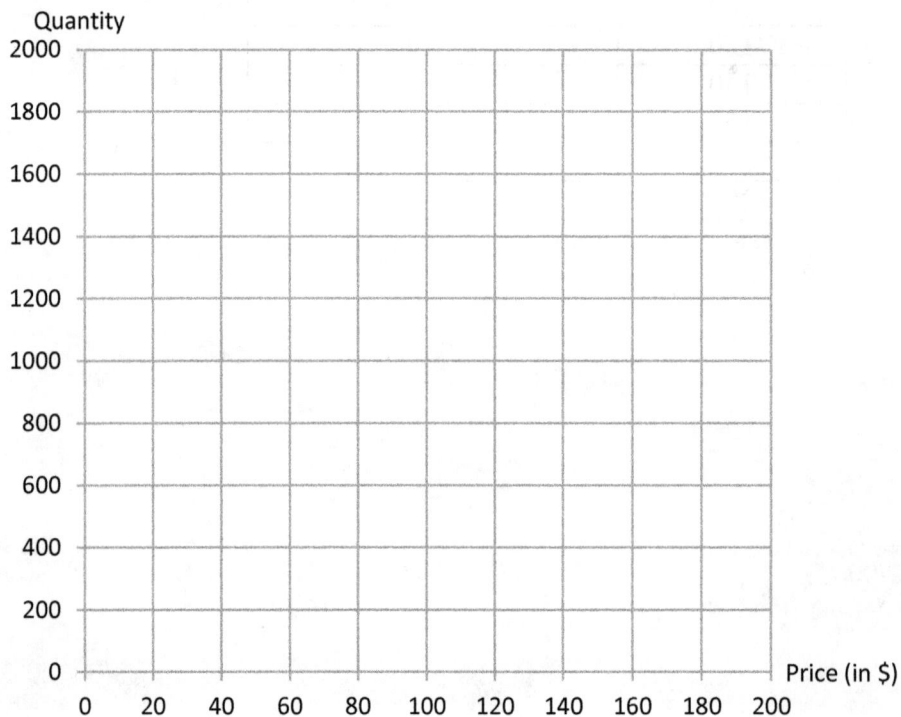

d. Market equilibrium between supply and demand occurs when the quantity supplied equals the quantity demanded. Place a mark on the graph in part *c* where the equilibrium point occurs. Use the graph to estimate the price and corresponding quantity that produce market equilibrium.

Estimated Market Equilibrium Price: _____

Estimated Market Equilibrium Quantity: _____

e. Find the market equilibrium price algebraically by substituting the expressions for demand *d* and supply *s* into the equilibrium equation below and then solving the resulting equation for price *p*. Record the market equilibrium price to the nearest penny. What quantity will be produced and/or supplied at this price? Round this quantity to the nearest unit.

$$\text{Equilibrium equation: } \quad d = s$$

Market Equilibrium Price: _____

Market Equilibrium Quantity: _____

f. At a price of \$180, what quantity is the producer willing to supply? How much would consumers be willing to purchase at that price? Explain using complete sentences.

g. In general, how do you think the market will react to a price that is higher than the market equilibrium price?

3. Conceptual Exercise—Vacation Rental Listings

A popular vacation rental listing website offers vacation property owners two fee options (A or B below) for listing and managing a rental property on the site.

Option A is a pay-per-rental plan where a property owner pays a 5% fee on the total rental cost charged to a renter, plus a 3% credit card fee on the total rental cost.

Option B is a subscription plan where a property owner pays a fixed $499 annual subscription fee for an unlimited number of rentals, plus a 3% credit card fee on the total rental cost.

A vacation property owner has a beach cottage that she rents for $250 per night. She would like to list her cottage for rent on this popular vacation rental listing website and wonders which fee plan she should choose.

a. Summarize the fee structure for these two options by completing the table shown below. Write "None" if a fee is not applicable.

Website Fee Plan	Fixed Subscription Fee	Pay-Per-Rental Fee	Credit Card Fee
Option A			
Option B			

b. If the beach cottage owner rents the cottage for 3 nights through the vacation rental listing website, how much will the owner owe in fees for this rental under Option A?

c. Let *x* represent the number of nights the beach cottage owner rents out the cottage over the course of a year. Write an algebraic expression for the total annual rental cost paid by renters of the beach cottage.

        Total annual rental cost = _____

d. Let *y* represent the total annual fees that the beach cottage owner pays for listing and renting it out through the website. Write an equation for each fee plan that gives the cottage's total annual fees, *y*, (in dollars) in terms of the annual number of nights, *x*, the cottage is rented out. Simplify each equation.

Option A: _____

Option B: _____

e. Solve the system of equations above to find the exact number of nights for which both fee plans charge exactly the same total annual fees. Decide whether to solve the system of equations by the substitution method or the addition method, and give reasons for this choice.

Solution Process:

Reasoning:

f. Make a recommendation to the owner of the beach cottage as to which fee plan she should choose, and explain your reasoning.

g. How would your recommendation change if the owner of the beach cottage could rent it for $500 per night instead of $250 per night?

4. Group Activity—Ship Course Analysis

From overhead photographs or satellite imagery of ships on the ocean, analysts can tell a lot about a ship's immediate course by looking at its wake. Assuming that two ships will maintain their present courses, it is possible to extend their paths, based on the wakes visible in the photograph. This can be used to find possible rendezvous points for the two ships. In this project, you will work with members of your group to investigate the courses of two ships shown in the figure.

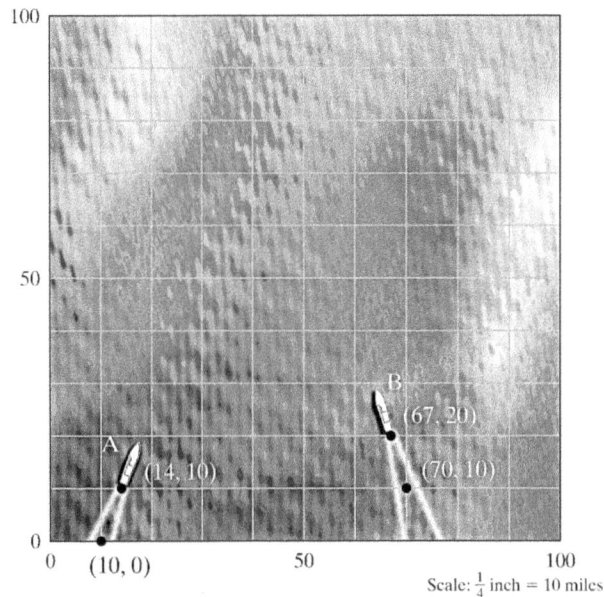

Scale: $\frac{1}{4}$ inch = 10 miles

a. Using each ship's wake as a guide, use a straightedge to extend the paths of the ships on the figure. Estimate the coordinates of the point of intersection of the ships' courses from the grid. If the ships continue in these courses, they could possibly rendezvous at the point of intersection of their paths.

b. What factor will govern whether or not the ships actually meet at the point found in part *a*?

c. Using the coordinates labeled on each ship's wake, find a linear equation that describes each path.

d. (Optional) Use a graphing calculator to graph both equations from part *c* in the same window. Use either the trace or intersect feature to estimate the point of intersection of the two paths. Compare this estimate to your estimate from part *a*.

e. Solve the system of two linear equations using one of the methods in this chapter. The solution is the point of intersection of the two paths. Compare your result to your estimate from part *a* (and part *d*, if applicable).

f. Plot the point of intersection you found in part *e* on the figure. Use the figure's scale (1/4 inch = 10 miles) to find each ship's distance in miles from this point of intersection by measuring from the bow (tip) of each ship with a ruler.

Ship A's distance from intersection point: _____

Ship B's distance from intersection point: _____

Now suppose that the speed of Ship A is r_1 and the speed of Ship B is r_2. Given the present positions and courses of the two ships, find a relationship between their speeds that would ensure a rendezvous.

1. Extension Exercise—Space Communications

In 2018, NASA announced plans to build a base of operations on Earth's moon. Once established, NASA will communicate with astronauts stationed there via radio signals that travel at the speed of light, which is known to be 186,000 miles per second.

a. Write the speed of light, as measured in miles per second, in scientific notation.

b. Use your answer from part *a* and the fact that 1 mile = 5.28×10^3 feet to convert the speed of light from miles per second to **feet per second.** Write the converted speed in scientific notation.

c. The distance between Earth and its moon is 1.2611544×10^9 feet, on average. Use the speed of light in feet per second (found in part *b*) and the distance formula $d = rt$ (where d is distance, r is rate, and t is time) to find how many seconds it will take a radio message traveling at the speed of light from NASA mission control on Earth to reach astronauts on the moon. Write your answer in scientific notation rounded to two decimal places.

d. Use your answer from part *a* and the fact that 1 mile ≈ 1.61 kilometers to convert the speed of light from miles per second to **kilometers per second.** Write the converted speed in scientific notation.

e. NASA intends to use the moon base as a staging area for space missions to other destinations in the solar system, such as Mars. When Earth and Mars are the closest, the distance between them is 5.46×10^7 kilometers. Use the speed of light in kilometers per second (found in part *d*) and the distance formula $d = rt$ to find how many seconds it will take a radio message traveling at the speed of light from NASA mission control on Earth to reach astronauts on Mars when Earth and Mars are at their closest. Write your answer in scientific notation rounded to two decimal places.

f. Use your answer from part *a* and the fact that 1 hour = 3.6×10^3 seconds to convert the speed of light from miles per second to **miles per hour.** Write the converted speed in scientific notation.

g. When Earth and Mars are farthest apart, the distance between them is 2.49×10^8 miles. Use the speed of light in miles per hour (found in part *f*) and the distance formula $d = rt$ to find how many hours it will take a radio message traveling at the speed of light from NASA mission control on Earth to reach astronauts on Mars when the planets are farthest apart. Write your answer in scientific notation rounded to two decimal places.

h. Explain why it might be challenging for astronauts on Mars to hold a radio conversation with mission control on Earth when Earth and Mars are farthest apart.

2. Exploration Activity—Deaths from Motor Vehicle Crashes

Each year the Insurance Institute for Highway Safety collects data on motor vehicle crash deaths in the United States, categorized by passenger vehicle occupants, pedestrians, motorcyclists, bicyclists, and large truck occupants.

According to data from the Insurance Institute for Highway Safety, a polynomial function that represents the annual number of deaths of **passenger vehicle occupants** over the period from 2006 through 2016 is $P(x) = 216x^2 - 2844x + 30,526$ where $x = 0, 1,..., 10$ represents the years 2006, 2007,..., 2016.

Also based on data from the Insurance Institute for Highway Safety, a polynomial function representing the annual number of deaths due to **all other types** of motor vehicle crashes (including pedestrians, motorcyclists, bicyclists, and large truck occupants) over the period from 2006 through 2016 is $A(x) = 68x^2 - 557x + 12,264$ where $x = 0, 1,..., 10$ represents the years 2006, 2007,..., 2016.

a. Use the polynomial functions to complete the table below. Find the number of passenger vehicle occupant deaths per year and the number of deaths due to all other types of motor vehicle crashes over the period of 2006 through 2016 by evaluating the polynomial functions $P(x)$ and $A(x)$ at the given values of x. Next, discuss how to use the data in the table to calculate the total number of motor vehicle crash deaths for each of the given years and then do so.

Year	x	Number of Deaths of Passenger Vehicle Occupants $P(x)$	Number of Deaths Due to All Other Types of Motor Vehicle Crashes $A(x)$	Total Number of Motor Vehicle Crash Deaths
2006	0			
2008	2			
2010	4			
2012	6			
2014	8			
2016	10			

b. Use the two given polynomials to find a new polynomial function that represents the **total number** of motor vehicle crash deaths per year and record it as $T(x)$ on the blank below. Evaluate this new polynomial function $T(x)$ for $x = 0, 2, 4, 6, 8$, and 10 and record your results in the table. Copy over your answers from the last column in part *a*.

New polynomial, $T(x) =$ _____

x	$T(x)$ **evaluated at given x-value**	**Data from last column in part a**
0		
2		
4		
6		
8		
10		

c. Compare the values in the second and third columns of the table from part *b*. What do you notice? What can you conclude?

3. Conceptual Exercise—Looking Back in Time

Light travels quickly but does not travel instantaneously. It can take years (or decades or even millennia) for light to travel from its source to a distant destination. If we see a distant star "burn out," it may have died years ago, but it has taken that long for the light of the explosion to reach Earth. For this reason, astronomers often say that when we see a star, we are *looking back in time.* The light we see from a star now is a snapshot of how the star looked when the light was first emitted years earlier.

In this exercise, we will explore the concept of a distance called a light year. Use a calculator to perform all necessary calculations along with the following information.

<u>Speed of light:</u>	186,000 or 1.86×10^5 miles per second
	300,000 or 3.0×10^5 kilometers per second
<u>Distance Formula:</u>	$d = rt,$ where d is distance, r is speed (or rate), and t is time

a. How many miles does light travel in 24 hours? Write your answer in scientific notation rounded to two decimal places.

b. How many kilometers does light travel in 24 hours? Write your answer in scientific notation rounded to two decimal places.

c. How many miles does light travel in 10 days? Write your answer in scientific notation rounded to two decimal places.

d. How many kilometers does light travel in 10 days? Write your answer in scientific notation rounded to two decimal places.

e. A "light year" is a unit of measure for length. One light year is defined as the distance that light travels in one Julian year, that is, 365.25 Earth days. How many miles are equivalent to 1 light year? Write your answer in scientific notation rounded to two decimal places. (*Note:* You may start your computation with your answer to part *a*.)

f. How many kilometers are equivalent to 1 light year? Write your answer in scientific notation rounded to two decimal places. (*Note:* You may start your computation with your answer to part *b*.)

g. An astronomer witnesses a star burning out and determines that it burned out about 100 years ago. How far was the star from Earth? Give your answer both in light years and in miles.

Star's distance from Earth in light years: _____

Star's distance from Earth in miles: _____

h. If a star is 25 light years away, how many miles is the star from Earth? Write your answer in scientific notation rounded to two decimal places. (*Note:* You may start your computation with your answer to part *e*.)

i. If we could travel at the speed of light, how long would it take us to travel to a planet that is 50 light years away?

4. Group Activity—Worldwide Internet Users

The number of people around the world who use the Internet continues to grow and grow. However, the worldwide population continues to increase as well, so it may be difficult to understand trends in the growth of Internet use with absolute numbers alone. In this activity, your group will explore the growth of worldwide Internet users as compared to the growth of worldwide population.

a. The number of worldwide Internet users (in millions) x years after the year 2000 is given by the polynomial function $P(x) = 8x^2 + 82x + 381$ for the years 2000 through 2017, according to data from Internet World Stats. Work with your group to complete the following table. The first line has been completed for you.

Year	x	evaluated at x	Number of worldwide Internet users written in **standard form**	Number of worldwide Internet users written in **scientific notation**
2000	0	381 million	381,000,000	3.81×10^8
2003				
2005				
2007				
2009				
2011				
2013				
2015				
2017				

b. The following table lists the approximate total worldwide population, according to the United States Census Bureau International Data Base, for each year given. Work with your group to complete the table by writing each population in standard form and in scientific notation.

Year	World population	World population written in **standard form**	World population written in **scientific notation**
2000	6.1 billion	6,100,000,000	6.1×10^9
2003	6.3 billion		
2005	6.5 billion		
2007	6.6 billion		
2009	6.8 billion		
2011	6.9 billion		
2013	7.1 billion		
2015	7.2 billion		
2017	7.4 billion		

c. Begin working on the table shown below by copying your results for the "Number of worldwide Internet users written in **scientific notation**" (last column of the table in part *a*) and for "World population written in **scientific notation**" (last column of the table in part *b*). Working with your group, use these values to complete the table by finding the percent of the world population that were Internet users in each year listed. Be sure to carry out your computations in scientific notation and check your work with a calculator. Round each percent to the nearest tenth of a percent.

Year	Number of worldwide Internet users written in **scientific notation**	World population written in **scientific notation**	Percent of world population that were Internet users
2000	3.81×10^8	6.1×10^9	6.2%
2003			
2005			
2007			
2009			
2011			
2013			
2015			
2017			

d. Describe any trends that you see in the data table in part *c*.

1. Extension Exercise—Maximizing the Area of a Pet Pen

Suppose you have 200 feet of fencing available and need to build a rectangular pet pen. You decide to build the pen along the back of the house so that you only need to fence in three sides (i.e., the house is the fourth side of the rectangle and does not need fencing). What dimensions will give a family pet the maximum area to live in?

a. Draw a sketch of the rectangular pen in its location against the back of the house. Label the side opposite the house as x and the two sides connecting the x-side to the house as y.

b. Write an equation for the length of the fencing needed using only variables x and y and the 200 feet of fencing that will enclose the three sides of the pen.

c. Solve for x in the equation from part *b*. Your equation for x should be written in terms of y.

d. The area of a rectangle is length times width, and so the formula $A = x \cdot y$ can be used to find the maximum area of the pen. Substitute the expression you obtained for x in part *c* into the area formula so that A is written only in terms of the variable y. Simplify your expression for A.

e. Find the two sets of dimensions, y and x, that can be used to build a rectangular pen with an area of 4800 square feet. Begin by substituting 4800 for A in your equation from part *d*. Use factoring to solve the resulting quadratic equation for y. (*Hint:* First rewrite the quadratic equation in standard form, i.e., $ax^2 + bx + c = 0$.) Then find the corresponding values for x by substituting each value of y into your expression for x from part *c*.

One set of pen dimensions that gives an area of 4800 sq. ft is $y =$ _____ ft and

$x =$ _____ ft. The other set of dimensions is $y =$ _____ ft and $x =$ _____ ft.

f. Explain how two *different* sets of dimensions can give the family pet the *same* area to play in.

g. Using the equation for *x* that you found in part *c*, complete the "Length *x*" column for the values of *y* given in the first column. Then use the area formula $A = \text{width} \times \text{length} = y \cdot x$ to complete the "Area *A*" column in the table.

Width *y* (feet)	Length *x* (feet)	Area *A* (sq. ft)
10		
20		
30		
40		
50		
60		
70		
80		
90		

h. Based on the table above, which dimensions will produce the maximum area of the pet pen?

i. There are other possible dimensions that could be used that are not listed in the table, such as 45 ft by 110 ft, or 55 ft by 90 ft. If you think that the dimensions from part *h* give the maximum possible area, explain why. Otherwise, find any dimensions that will produce a greater area for the pet pen.

2. Exploration Activity—More Space, Less Zeus

Suppose you want to plant a rectangular-shaped vegetable garden, but you need to protect it from your neighbor's dog, Zeus. Although Zeus is kept in a pen, his owners often forget to latch the gate securely. You have not seen the dog escape yet but plan to use 18 meters of wire fencing to enclose the perimeter of your garden, just in case. Assuming that you use all 18 meters of fencing, what dimensions will give you the maximum amount of area for planting?

a. Draw a picture of your garden labeling the length y and the width x.

b. Substitute the 18 meters of wiring for Perimeter in the equation below and then isolate one of the variables so that your resulting equation is in terms of one variable only.

$$\text{Perimeter} = 2(x + y)$$

c. Substitute the expression for the variable that you isolated in part *b* into the area equation below.

$$A = x \cdot y$$

d. Find the dimensions that are needed to build a rectangular garden with an area of 8 square meters. Start by substituting 8 for A in the equation obtained in part *c*. Solve the resulting quadratic equation by factoring. (*Hint:* First put the equation in standard form, i.e., $ax^2 + bx + c = 0$.)

e. Repeat part *d* for gardens with the following areas:
- 14 square meters
- 18 square meters
- 20 square meters

f. If you use only 18 meters of wire fencing, can you build a garden with an area greater than 20 square meters? Complete the following table to support your answer.

Length (meters)	Width (meters)	Perimeter (meters)	Area (sq. m)
4.1			
4.2			
4.3			
4.4			
4.5			
4.6			
4.7			
4.8			
4.9			

g. What is the maximum possible area of the garden? Use the table in part *f.*

h. What dimensions from the table give the maximum area for planting a garden?

i. Based on the calculations you have made for the garden, what special type of rectangular shape will produce the maximum area for a fixed perimeter?

3. Conceptual Exercise—A Landscaping Design

Suppose you are a landscaper. You are landscaping a public park and have just put in a rectangular flower bed measuring 8 feet by 12 feet. You would also like to surround the bed with a decorative floral border consisting of low-growing, spreading plants. Each plant will cover approximately 1 square foot of ground when mature, and you have 224 plants to use. For the sake of aesthetics, the border should be the same width all the way around the 8 ft × 12 ft flower bed. Ultimately, you will need to find how wide of a strip of ground you should prepare around the flower bed for your decorative floral border.

Because the width of the decorative floral border is unknown at this point, let's call it *x*. The figure below shows the dimensions of the flower bed and surrounding floral border.

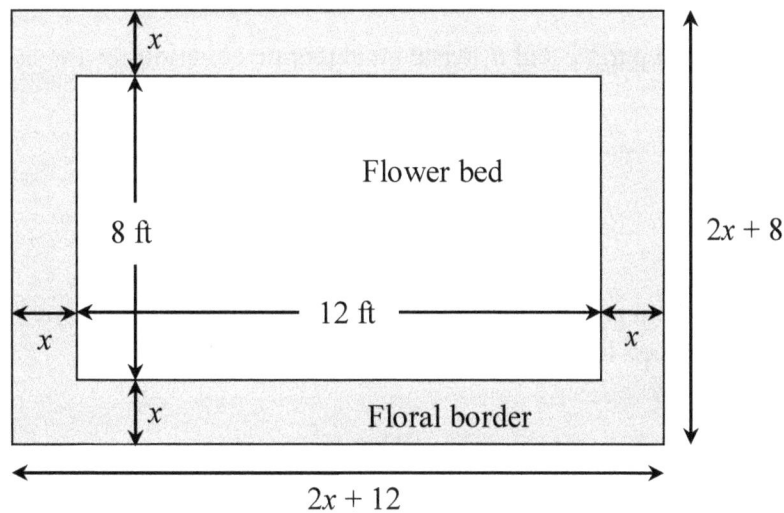

a. Find the area of the rectangular flower bed by itself.

b. Use the dimensions given in the figure to write and simplify an algebraic expression for the total combined area of the flower bed and surrounding floral border.

c. Use subtraction and your results from parts *a* and *b* to find a simplified algebraic expression for the area of *just the floral border* (not including the area of the rectangular flower bed).

d. Knowing that each of the 224 spreading plants that you plan to use for the floral border will eventually cover 1 square foot of ground, what is the total area of the ground that you will need to prepare for the floral border?

e. Using your results from parts *c* and *d*, write an algebraic equation for the area of the floral border.

f. Solve your equation from part *e*. How wide of a strip of ground should you prepare around the rectangular flower bed for the floral border? Explain.

4. Group Activity—The 10-Meter Dive

Suppose a diver jumps from a platform that is 10 meters above the water surface at an initial velocity of 5 meters per second. The height, h (in meters), of the diver's position above the water at any time t (in seconds) is given by the following equation,

$$h = -5t^2 + 5t + 10.$$

a. How many seconds, after jumping, does it take for the diver to enter the water? To answer this question, let $h = 0$ and solve by factoring.

b. Use $h = -5t^2 + 5t + 10$ and complete the table to see how the height of the diver's position changes from 0 to 2.5 seconds. Round to the nearest hundredth of a meter, if necessary.

Time, t (seconds)	Height, h (meters)
0	
.25	
.50	
.75	
1.00	
1.25	
1.50	
1.75	
2.00	
2.25	
2.50	

c. Explain how the height of the diver's position changed over the interval of time in the table.

   d. How many seconds after the jump did the diver reach the maximum height? What is that maximum height? Write your answer in a complete sentence.

1. Extension Exercise—Financing Jamille's Car

To determine the monthly payment for a car loan, use the formula $L \cdot \dfrac{r(1+r)^n}{(1+r)^n - 1}$ where L is the

amount financed, r is the monthly interest rate (annual rate $\div$ 12), and n is the number of months of the loan. For example, to compute the monthly payment for a $12,000 car loan, financed for 3 years at an 8.5% annual interest rate, assign values to each variable as follows and use a calculator to perform the calculations.

- $L = 12{,}000$
- $n = 36$ (3 years at 12 months per year)
- $r = 0.0071$ (found by dividing 0.085 by 12, rounded to the nearest ten-thousandth)

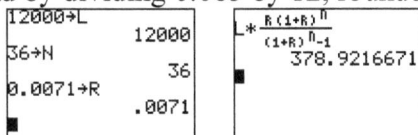

Jamille has found the car of his dreams and has negotiated a price of $15,798 after trading in his old car. He will finance the full negotiated price of the new car. Here are his financing options for a 4-year car loan:

- Option 1: 8.9% manufacturer's loan with a $1000 price rebate at signing
- Option 2: 7.7% bank loan with no rebate
- Option 3: 9.1% home equity loan with no rebate

The advantage of taking out a home equity loan is that the interest paid on the loan is often tax deductible. If Jamille chooses Option 1, then he will deduct the rebate from the negotiated price and finance the remaining amount.

a. Find the monthly payment for each of Jamille's three options using the monthly payment formula given above. In the third column of the table below, show the values you used for n and r. **Round r to four decimal places.** In the fourth column, record the monthly payment you computed to the nearest cent. (Remember that this is a 4-year loan.)

Financing Option	Amount Financed	Values Substituted into Formula	Monthly Payment
1. 8.9% loan with a $1000 rebate	$L = \$14{,}798$	$n =$ $r =$	
2. 7.7% loan with no rebate	$L = \$15{,}798$	$n =$ $r =$	
3. 9.1% home equity loan with no rebate	$L = \$15{,}798$	$n =$ $r =$	

b. For each payment option, write the total amount A that Jamille will pay after making 48 monthly payments.

   Option 1, 8.9% loan with a $1000 rebate: $A =$ _____

   Option 2, 7.7% loan with no rebate: $A =$ _____

   Option 3, 9.1% home equity loan with no rebate: $A =$ _____

c. The total interest paid is the total of the payments A minus the original amount financed L. Fill in the following table to compare Jamille's cost of buying the car using the three different financing options. If necessary, round amounts to the nearest cent.

	Option 1 8.9% loan with a $1000 rebate	**Option 2** 7.7% loan with no rebate	**Option 3** 9.1% home equity loan with no rebate
Total Payments A (From part *b*)			
Amount Financed L (From table on previous page)			
Total Interest Paid $(A-L)$			
Tax Savings S (From taking Home Equity Loan)	$0	$0	(24% of total interest paid, from line above)
Cost after Tax Savings $(A-S)$			

d. Based on the data from part *c*, explain to Jamille which financing option would be the best choice. Include data to support your conclusions.

2. Exploration Activity—Children's Doses of Medicine

Doctors commonly use dose formulas for prescribing medicines to children. These dose formulas describe an approximate relationship only. Young's Rule and Cowling's Rule (two dose formulas) both relate a child's age A in years and an adult dose D of medication to the proper child's dose C. The formulas are most accurate when used for children between the ages of 2 and 13.

Young's Rule: $C = \dfrac{DA}{A+12}$ **Cowling's Rule:** $C = \dfrac{D(A+1)}{24}$

a. Let the adult dose $D = 1000$ mg. Fill in the table below, which compares the doses predicted by both formulas for children of ages $A = 2, 3, 4, \ldots, 13$. If necessary, round doses to the nearest tenth.

Age of Child	Child's Dose by Young's Rule	Child's Dose by Cowling's Rule
2		
3		
4		
5		
6		
7		
8		
9		
10		
11		
12		
13		

b. Use your table from part *a* to decide whether either formula will consistently predict a larger dose than the other. If so, which formula does? If not, is there an age at which the doses predicted by one formula become greater than the doses predicted by the other formula? If this is true, estimate that age. Explain your reasoning.

c. Would your conclusion in part *b* remain the same if the related adult dose *D* was something other than 1000 mg? Explain how you arrived at your conclusion.

3. Conceptual Exercise—Epigram of Diophantus

One of the great algebraists of ancient times was a man named Diophantus. Little is known of his life other than that he lived and worked in Alexandria. Some historians believe he lived during the first century of the Common Era. The only clue to his personal life is the following epigram found in a collection called the Palatine Anthology.

> "This tomb holds Diophantus. Ah, how great a marvel! The tomb tells scientifically the measure of his life. God granted him to be a boy for the sixth part of his life, and adding a twelfth part to this, He clothed his cheeks with down; He lit him the light of wedlock after a seventh part, and five years after his marriage He granted him a son. Alas! Late-born wretched child; after attaining the measure of half his father's life, chill Fate took him. After consoling his grief by this science of numbers for four years he ended his life."

The question is: how old was Diophantus when he died? To begin, let x represent that age. If we add up the various parts of his life as described, we should get the age at which he died.

Parts of Diophantus's life $\begin{cases} \dfrac{x}{6} + \dfrac{x}{12} \text{ is the time of his youth} \\[2mm] \dfrac{x}{7} \text{ is the time between his youth and when he married} \\[2mm] 5 \text{ years is the time between his marriage and the birth of his son} \\[2mm] \dfrac{x}{2} \text{ is the time Diophantus had with his son} \\[2mm] 4 \text{ years is the time between his son's death and his own} \end{cases}$

a. Write an equation that can be solved to find how old Diophantus was when he died.

b. Solve the epigram; that is, find the age at which Diophantus died.

c. How old was Diophantus when his son was born?

d. From the epigram, we know that Diophantus outlived his son. How long did his son live?

e. Solve the following epigram:

   I was four when my mother packed my lunch and sent me off to school. Half my
   life was spent in school and another sixth was spent on a farm. Alas, hard times
   befell me. My crops and cattle fared poorly and my land was sold. I returned to
   school for 3 years and have spent one-tenth of my life teaching. How old am I?

f. Write an epigram describing your life. Be sure that none of the time periods in your epigram
   overlap. Exchange epigrams with another student in your class to solve and check.

4. Group Activity—Nutrition

In order to help consumers make informed food choices, nutrition facts labels for packaged foods give many important details about the nutrients provided by the item. Each label gives information based on a standard portion size. However, many consumers eat an amount that is quite different from the standard portion size. Proportions are a very useful tool in analyzing the nutrition of nonstandard portion sizes, helping consumers find out exactly what they are eating and how that relates to their nutritional needs.

Breakfast Cereal

Nutrition Facts
Serving Size 1.25 cup (59g)

Amount Per Serving

Calories 200	Calories from Fat 10

	% Daily Values*
Total Fat 1g	2%
Saturated Fat 0g	0%
Trans Fat 0g	
Polyunsaturated Fat 0.5g	
Cholesterol 0mg	0%
Potassium 230mg	7%
Sodium 0mg	0%
Total Carbohydrate 48g	16%
Dietary Fiber 9g	36%
Sugars 0g	
Protein 6g	12%

Calcium 2%	●	Iron 15%
Thiamin 8%	●	Riboflavin 2%
Niacin 15%	●	Vitamin B6 2%
Folate 4%	●	Phosphorus 20%
Magnesium 20%	●	Zinc 10%
Copper 10%		

*Percent Daily Values are based on a 2,000 calorie diet. Your Daily Values may be higher or lower depending on your calorie needs.

	Calories	2,000	2,500
Total Fat	Less than	65g	80g
Sat Fat	Less than	20g	25g
Cholesterol	Less than	300mg	300mg
Sodium	Less than	2400mg	2400mg
Total Carbohydrate		300g	375g
Dietary Fiber		25g	30g

a. Every morning Nami pours herself 75 grams of cereal for breakfast. The nutrition facts for her breakfast cereal are listed at the right. How many calories does her daily serving of cereal deliver? Round to the nearest whole calorie.

b. How much dietary fiber does Nami get from her 75-gram serving of cereal? Round to the nearest tenth of a gram. If her goal is to get 25 grams of fiber each day, how much more fiber does Nami need in order to reach her goal?

c. How many grams of protein does Nami receive from her 75-gram serving of cereal? Round to the nearest tenth of a gram.

d. Mateo's plain yogurt has the nutrition label shown below. If he has a 145-gram serving of yogurt with his lunch, how much potassium does it provide? Round to the nearest whole milligram.

Nutrition Facts

Serving Size 1 cup (227g)
Servings Per Container 4

Calories 120
  Calories from Fat 20

Amount Per Serving	% Daily Values*	Amount Per Serving	% Daily Values*
Total Fat 2g	3%	**Sodium** 140mg	6%
Saturated Fat 1.5g	8%	**Total Carbohydrate** 15g	5%
Trans Fat 0g		Dietary Fiber 0g	0%
Cholesterol 15mg	5%	Sugars 15g	
Potassium 480mg	14%	**Protein** 10g	20%

Vitamin A 2%   •   Calcium 35%   •   Vitamin D 25%

*Percent Daily Values are based on a 2,000 calorie diet. Your Daily Values may be higher or lower depending on your calorie needs.

	Calories	2,000	2,500
Total Fat	Less than	65g	80g
Sat Fat	Less than	20g	25g
Cholesterol	Less than	300mg	300mg
Sodium	Less than	2400mg	2400mg
Total Carbs		300g	375g
Dietary Fiber		25g	30g

Proportions are also useful for finding the portion size of a certain food that is required to provide a specific nutrient or calorie level.

e. Sophie is concerned that she isn't getting enough fiber in her diet. After carefully tracking her fiber intake for the day, she discovers that she is 5 grams short of her fiber goal of 25 grams for the day. If Sophie snacks on some dried fruit before bed (see the nutrition facts label at the right), how much fruit (in grams) should she eat to ensure that she reaches her fiber goal for the day? Round to the nearest whole gram.

Dried Fruit

Nutrition Facts

Serving Size 0.25 cup (40g)
Servings Per Container 31

Amount Per Serving

Calories 100

	% Daily Values*
Total Fat 0g	0%
Saturated Fat 0g	0%
Trans Fat 0g	
Cholesterol 0mg	0%
Potassium 320mg	9%
Sodium 0mg	0%
Total Carbohydrate 23g	8%
Dietary Fiber 3g	12%
Sugars 19g	
Protein 1g	2%

Vitamin A 15%	•	Calcium 2%
Iron 20%	•	Vitamin E 20%
Vitamin B6 20%	•	Vitamin B12 20%

*Percent Daily Values are based on a 2,000 calorie diet. Your Daily Values may be higher or lower depending on your calorie needs.

	Calories	2,000	2,500
Total Fat	Less than	65g	80g
Sat Fat	Less than	20g	25g
Cholesterol	Less than	300mg	300mg
Sodium	Less than	2400mg	2400mg
Total Carbohydrate		300g	375g
Dietary Fiber		25g	30g

f. The Chopra family buys candy-coated chocolate pieces in bulk and re-packages them at home in 100-calorie portions. Use the nutrition facts shown below to find how many grams of candy would make up a 100-calorie portion. If each candy-coated chocolate piece weighs 0.85 gram on average, how many individual candy pieces are in an average 100-calorie portion? Round to the nearest whole number.

Nutrition Facts	Amount Per Serving		% Daily Values*	Amount Per Serving		% Daily Values*	*Percent Daily Values are based on a 2,000 calorie diet. Your Daily Values may be higher or lower depending on your calorie needs.			
Serving Size 1.5 oz (42g)	**Total Fat** 9g		14%	**Total Carbohydrate** 30g		10%				
	Saturated Fat 6g		30%	Dietary Fiber 1g		4%		Calories	2,000	2,500
	Trans Fat 0g			Sugars 27g			Total Fat	Less than	65g	80g
Servings Per Container 50							Sat Fat	Less than	20g	25g
Calories 210	**Cholesterol** 5mg		2%	**Protein** 2g		4%	Cholesterol	Less than	300mg	300mg
Calories from Fat 80	**Sodium** 25mg		1%				Sodium	Less than	2400mg	2400mg
							Total Carbs		300g	375g
	Calcium 4%	•	Iron 2%	•	Riboflavin 2%		Dietary Fiber		25g	30g

Added sugar in one's diet is linked to increased risk for diabetes, high cholesterol, inflammation, and high blood pressure. The American Heart Association recommends that women restrict added sugars in their diets to no more than 6 teaspoons (or 24 grams) per day and that men restrict added sugars in their diets to no more than 9 teaspoons (or 36 grams) per day.

g. Diego has been carefully tracking his added sugar intake for the day, and by the end of the day he finds that he has already ingested 27 grams of added sugar. He would like to treat himself with a serving of the candy-coated chocolate pieces described in part *f.* Assuming that all of the candies' reported sugar is added sugar, what is the maximum number of grams of candy that Diego could eat without going over the recommended daily limit for added sugar? If each candy-coated chocolate piece weighs 0.85 gram on average, what is the maximum whole number of candy pieces that he could eat without going over his limit for added sugar?

h. Use one of the nutrition facts labels in this activity to write your own nutrition exercise. Exchange your nutrition exercise with another member of your group and solve one another's exercises.

1. **Extension Exercise—Hydrology**

Hydrology is the study of water: where it occurs on our planet, how it moves, and what qualities it has. Civil and environmental engineers must understand the water cycle and must consider various hydrological factors such as rainfall intensity (the depth of rain that falls during a certain time period) and runoff rates (the flow of water in volume per unit time) when designing water treatment facilities and storm sewer systems.

The peak runoff rate Q (in cubic feet per second) of a plot of land is jointly proportional to the rainfall intensity I of a storm (in inches per hour) and the area A of the land plot (in acres) under consideration. The constant of proportionality c in this relationship is known as the *runoff coefficient*, and its value depends on the type of surface in the drainage area and whether that area is flat or has a slope.

a. Carefully read the second paragraph above and write an equation for peak runoff rate Q. Then solve the equation for the runoff coefficient c.

b. A 2-acre drainage area consists of flat pavement. During a particular rainstorm, the rainfall intensity was 0.20 inches per hour and the peak runoff rate for this drainage area was 0.36 cubic feet per second. What is the runoff coefficient for flat pavement?

c. When it rains at a rate of 0.10 inches per hour on a relatively flat meadow encompassing 5 acres, a peak runoff rate of 0.125 cubic feet per second is produced. What is the runoff coefficient for flat meadow land?

d. A farmer owns a flat 12-acre cultivated field. If the peak runoff rate for this cultivated field is 1.8 cubic feet per second when rainfall intensity is 0.30 inches per hour, find the peak runoff rate for this same 12-acre cultivated field when rainfall intensity drops to 0.15 inches per hour.

e. A 20-acre plot of flat forest land exhibits a peak runoff rate of 0.50 cubic feet per second when rain falls at a rate of 0.25 inches per hour. Find the peak runoff rate for a similar flat forest with a 16-acre drainage area when rainfall intensity is 0.18 inches per hour.

f. Review your work in parts *b-e* to complete the table listing runoff coefficients for various drainage area surfaces.

Drainage Area Surface	Runoff Coefficient, *c*
Flat forest	
Flat meadow	
Flat cultivated field	
Flat pavement	

g. A city worker is collecting data at the site of a flat 4-acre paved parking lot. A sudden rainstorm hits while she works on site. She uses a flow meter to measure runoff rates from the parking lot throughout the storm. Scanning her data, she identifies 0.80 cubic feet per second as the peak runoff rate during the storm. Use what you know about the situation to estimate the rainfall intensity of the sudden storm. Round your answer to the nearest hundredth.

2. Exploration Activity—Earthquake Prediction

The San Andreas Fault in California is a break in the earth's crust where many earthquakes have occurred. Parkfield is an area in central California that seismologists have studied to predict future earthquakes. The following table shows the years since 1857 in which the study center in Parkfield has measured moderate earthquakes of about magnitude 6.0. Read the quake number as first, second, third, and so on.

Quake Number x	1	2	3	4	5	6
Year y	1857	1881	1901	1922	1934	1966

(*Source:* U.S. Geological Survey)

a. Plot the ordered pairs of data as points on the rectangular coordinate system below. Then use a straight edge to draw a line that comes closest to all of the points, but does not necessarily go through any of them. This line is called your best-fit line.

Year y

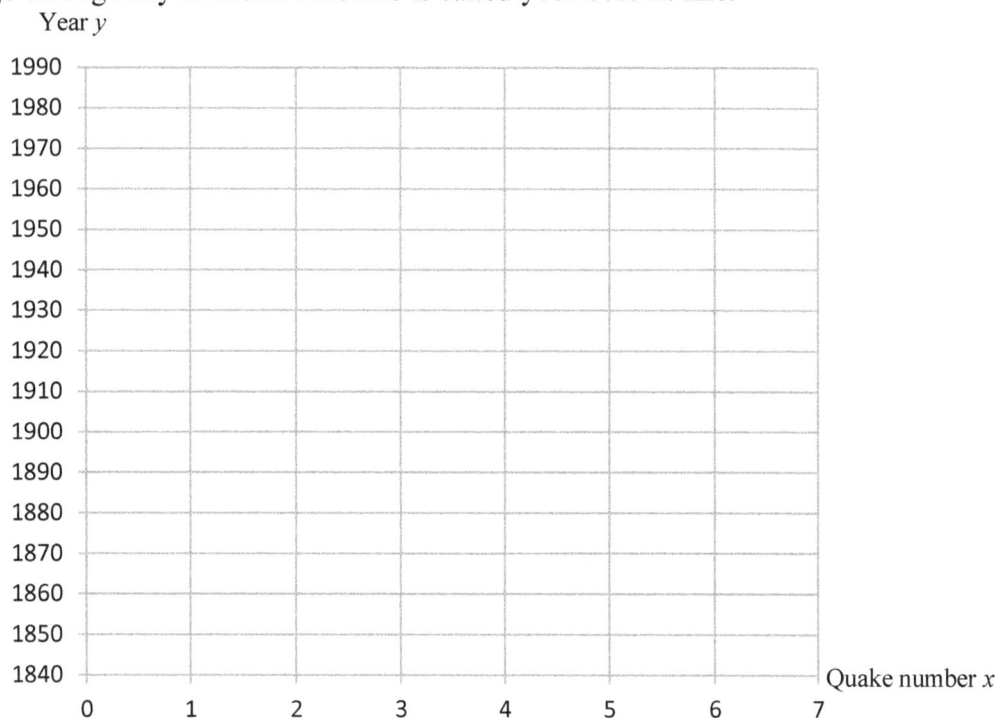

b. Find out how much time passed between earthquakes of magnitude 6.0 by finding each successive difference in y-values, denoted as Change in Time. The first value in that column has been computed for you.

Quake Number x	Year y	Change in Time (in years)
1	1857	
2	1881	$1881 - 1857 = 24$
3	1901	
4	1922	
5	1934	
6	1966	

c. Find the average amount of time between these moderate earthquakes by computing the average of the values in the third column of the table in part *b*.

d. Use your graph from part *a* and the average amount of time between moderate earthquakes computed in part *c* to predict when the seventh earthquake should have occurred. Explain your answer in complete sentences.

e. Estimate the coordinates of two points that fall on your best-fit line. Use these points to find the slope of your line. Show all your work.

f. What does the value of the slope of your best-fit line mean in terms of the problem situation?

g. Use the point-slope form of a linear equation $y - y_1 = m(x - x_1)$ to find the equation of your best-fit line. Then write the equation using function notation.

h. Find $f(7)$ and explain what this means in the context of the Parkfield earthquake situation. How does this answer compare to your prediction in part *d*?

3. Conceptual Exercise—Target Heart Rate

According to the American Heart Association, your target heart rate should stay within 50 to 85 percent of your maximum heart rate during exercise. The theory is to begin near 50% of your maximum heart rate and gradually work your way up to no more than 85% of your maximum heart rate as you progress through a fitness program. The table displays sample average maximum heart rate M in beats per minute (bpm) that corresponds to each age x from 20 years to 60 years in steps of 5 years.

Age x (in years)	Average Maximum Heart Rate M (in bpm)
20	200
25	195
30	190
35	185
40	180
45	175
50	170
55	165
60	160

(*Source:* American Heart Association)

a. Describe the relationship between maximum heart rate and age by completing the sentence below.

As age increases by 5 years, average maximum heart rate _____

_____

b. For each constant change in the input variable x, there is a constant change in output variable M. What type of function exhibits this type of behavior?

c. The relation between Age and Average Maximum Heart Rate above describes a function. For the domain [20, 60], state the range of the function.

d. Find the slope of this function. To do so, choose two ordered paris (x, M) from the table.

$$m = \frac{\text{Change in } M}{\text{Change in } x} =$$

e. Use the table on the previous page to increase the number of ordered pairs so that it includes $x = 0, 5, 10,$ and 15 as ages in the first column. Find the average maximum heart rate M for each of these ages.

Age x	Average Maximum Heart Rate M
0	
5	
10	
15	

f. Write an equation that gives average maximum heart rate M as a function of age x.
Hint: Use the slope-intercept form $f(x) = mx + b$ where the constant b is the output for the input 0 found in part *e*.

$M(x) = $ _____

g. Suppose Steve, age 41, wants to start an exercise program. Use the function from part f to find his average maximum heart rate.

$M(41) = $ _____

h. If Jimena has an average maximum heart rate of 147 bpm, find her age by setting the function from part f equal to 147 and then solving for x.

i. Answer the following questions to calculate Steve's **target heart rate zone** from 50% to 85% of his average maximum heart rate. Round heart rates to the nearest whole number.
- What is 50% of Steve's average maximum heart rate?
- What is 85% of Steve's average maximum heart rate?
- Fill in the blanks: Steve's target heart rate zone is from _____ bpm to _____ bpm.

j. Complete the following table giving the target heart rate zone for the listed ages. Round heart rates to the nearest whole number.

Age (Years)	Target Heart Rate Zone 50 – 85% (in bpm)
20	
25	
30	
35	
40	
45	
50	
55	
60	

4. Group Activity—Global Warming

About 87% of all human-produced emissions of carbon dioxide come from burning fossil fuels such as coal, oil, and gas. Currently oceans and forests absorb only about 50% of carbon emissions caused from burning fossil fuels. The rest remains in the atmosphere. As carbon dioxide accumulates in the atmosphere, it traps the sun's heat. Scientists believe that increasing concentrations of carbon dioxide and other greenhouse gases are causing global warming.

The table below gives the annual atmospheric concentrations of carbon dioxide in parts per million (ppm) for the period from 2000 to 2018.

Time (year)	Years after 2000 x	Carbon Dioxide Concentration y (ppm)	Annual Change
2000	0	369.55	
2001	1	371.14	$371.14 - 369.55 = 1.59$
2002		373.28	
2003		375.80	
2004		377.52	
2005		379.80	
2006		381.90	
2007		383.79	
2008		385.60	
2009		387.43	
2010		389.90	
2011		391.65	
2012		393.85	
2013		396.52	
2014		398.65	
2015		400.83	
2016		404.24	
2017		406.55	
2018		408.52	

(*Source:* NOAA Earth System Research Laboratory)

a. Let the variable x represent the number of years after the year 2000, so that $x = 0$ represents the year 2000. Complete the column for x in the table. Next, use a calculator to find the annual change in carbon dioxide concentration from the previous year, starting with 2001. Record your results in the column labeled "Annual Change." The first row has been completed for you.

b. Use a calculator to find the *average* annual change in carbon dioxide concentration for the given time period. Write a sentence interpreting your result.

c. Create a scatter diagram of the paired data given in the table as ordered pairs (x, y). Describe any pattern you see in the graph.

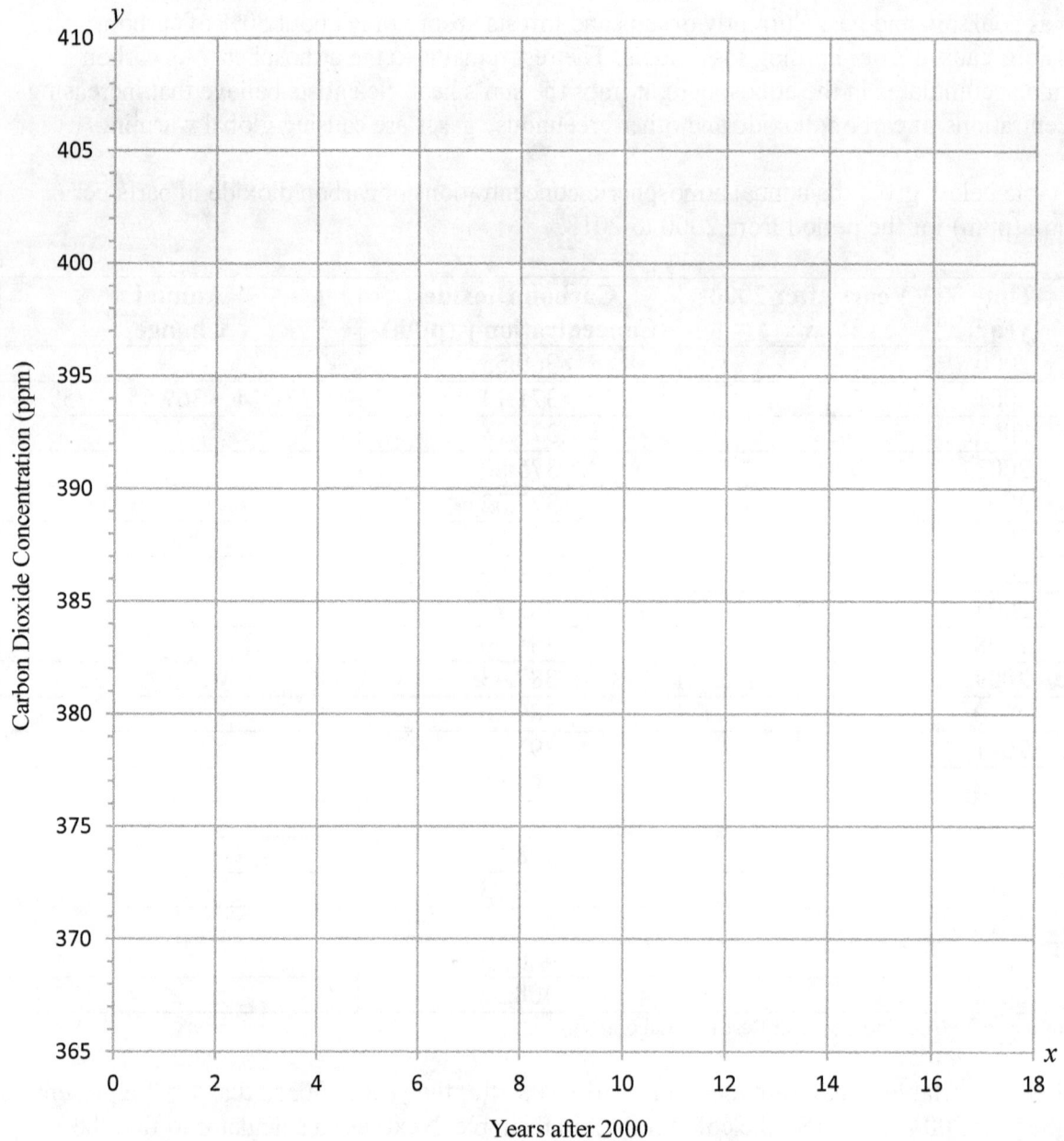

d. To develop a linear model of the form $f(x) = mx + b$, use the average annual change from part *b* as the slope *m*. To find the *y*-intercept *b*, use the table and find the output *y* when the input *x* is zero.

Slope *m* = _____   *y*-intercept *b* = _____

e. Use the slope *m* and *y*-intercept *b* to find a linear function $f(x) = mx + b$ to model carbon dioxide concentration *f* as a function of time *x* in years after 2000.

Linear function $f(x)$ = _____

f. Graph your linear function from part *e* on the scatter diagram you created in part *c*. Observe how close the line comes to all of the data points. Does your linear function appear to be a good fit for the given data? Explain.

g. Use your linear function from part *e* to predict the atmospheric concentration of carbon dioxide in the year 2025. In other words, find $f(25)$.

h. Use your linear function from part *e* to predict the atmospheric concentration of carbon dioxide in the year 2060. (Hint: See part *g*.)

i. The 2018 carbon dioxide concentration was about 409 parts per million (ppm). Assuming that the linear trend continues into the future, when will carbon dioxide concentration reach a level that is double the 2018 level? Show all your work, and explain your result in complete sentences.

1. Extension Exercise—Set Operations

Let *L* be the set of U.S. states that border the Great Lakes. Then

L = {Illinois, Indiana, Michigan, Minnesota, New York, Ohio, Pennsylvania, Wisconsin}.

Let *M* be the set of U.S. states that begin with the letter M. Then

M = {Maine, Maryland, Massachusetts, Michigan, Minnesota, Mississippi, Missouri, Montana}.

Let *N* be the set of U.S. states that begin with the letter N. Then

N = {Nebraska, Nevada, New Hampshire, New Jersey, New Mexico, New York, North Carolina, North Dakota}.

Finally, let *S* be the set of U.S. states that, according to the Insurance Institute for Highway Safety, had a maximum posted daytime speed limit of 65 mph on rural interstates as of May 2019. Then

S = {Alaska, Connecticut, Delaware, Massachusetts, New Jersey, New York, Rhode Island, Vermont}.

a. List the elements of the set $M \cup N$. Describe the elements of the union of these two sets in a complete sentence.

b. List the elements of the set $L \cup M$. Describe the elements of the union of these two sets in a complete sentence.

c. List the elements of the set $N \cap S$. Describe the elements of the intersection of these two sets in a complete sentence.

d. List the elements of the set $L \cap S$. Describe the elements of the intersection of these two sets in a complete sentence.

e. Find $M \cap N$. Explain the meaning of this set.

The **distributive law for sets** states that for any sets A, B, and C,

$$A \cap (B \cup C) = (A \cap B) \cup (A \cap C) \quad \text{and} \quad A \cup (B \cap C) = (A \cup B) \cap (A \cup C)$$

f. Check the first form of the distributive law for sets by finding $L \cap (M \cup N)$ and $(L \cap M) \cup (L \cap N)$. Then verify that the two sets are the same. (*Note:* Just as in order of operations, make sure you perform the operation within parentheses first.)

g. Check the second form of the distributive law for sets by finding $S \cup (M \cap N)$ and $(S \cup M) \cap (S \cup N)$. Then verify that the two sets are the same.

2. Exploration Activity—Final Grades

Dr. Alvarez teaches algebra at Big State University. She computes the final grade for each of her algebra students using a weighted average based on the following weights.

Final Grade Weights	
Quizzes	10%
Homework	20%
Test 1	20%
Test 2	20%
Final Exam	30%

For instance, a student's score on the final exam accounts for 30% of the student's final grade for the course, a student's average quiz score accounts for 10% of the student's final grade, and so on. Notice that the total of the percent column above is 100%.

Once Dr. Alvarez has computed a final grade for each student, she assigns a letter grade according to the following grading scale.

Final Grading Scale	
A	93% to 100%
A–	90% to 92%
B+	87% to 89%
B	83% to 86%
B–	80% to 82%
C+	77% to 79%
C	73% to 76%
C–	70% to 72%
D+	67% to 69%
D	65% to 66%
F	0% to 64%

Let's assign some variables for a student's scores. Let:

Q = average score on Quizzes $\qquad$ T_2 = score on Test 2

H = average score on Homework $\qquad$ F = score on Final Exam

T_1 = score on Test 1

For each category, scores are reported as percents.

a. Using the variables and Final Grade Weights shown above, write an algebraic expression for the portion of a student's final grade attributed to quizzes.

b. Using the variables and Final Grade Weights shown above, write an algebraic expression for the portion of a student's final grade attributed to homework.

c. Using the variables and Final Grade Weights shown on the previous page, write an algebraic expression for the portion of the student's final grade attributed to test 1.

d. Using the variables and Final Grade Weights shown on the previous page, write an algebraic expression for the portion of the student's final grade attributed to test 2.

e. Using the variables and Final Grade Weights shown on the previous page, write an algebraic expression for the portion of the student's final grade attributed to the final exam.

f. Using your answers to parts *a* through *e*, write an algebraic expression that Dr. Alvarez can use to calculate a student's final grade.

g. Now, using your expression from part *f* and the Final Grading Scale shown on the previous page, write a compound inequality that represents the range of final grades that will be assigned the letter grade B+.

h. Similarly, write a compound inequality that represents the range of final grades that will be assigned the letter grade C.

i. Braxton is a student in Dr. Alvarez's algebra class. Just before taking the final exam, she had scores of 78% on quizzes, 92% on homework, 75% on test 1, and 88% on test 2. In what range can Braxton score on her final exam to be assured of receiving a letter grade of B in the course?

j. Diego is another student in Dr. Alvarez's algebra class. Just before taking the final exam, he had scores of 66% on quizzes, 75% on homework, 76% on test 1, and 72% on test 2. In what range can Diego score on his final exam to be assured of receiving a letter grade in the C range (that is, either C–, C, or C+) in the course?

k. Just before taking Dr. Alvarez's final exam in algebra, Carter had scores of 88% on quizzes, 95% on homework, 86% on test 1, and 87% on test 2. Use a compound inequality to decide if it is possible for Carter's score on the final to guarantee an A in the course. Explain your reasoning.

3. Conceptual Exercise—Tolerances

In manufacturing processes, it is often impossible to make an item to an exact specification. When items are manufactured repeatedly on a production line, there is almost always variation in the resulting dimensions—such as length, width, diameter, weight, thickness, volume, and so on—of the many items. Manufacturing engineers usually specify **tolerances** for manufactured items that describe how much variation in dimension is acceptable for a particular item. For example, if the target diameter of a manufactured bolt is 50 mm and it is deemed "acceptable" if the actual diameter is within 0.25 mm of the target diameter, then any bolt with a diameter between 49.75 mm and 50.25 mm, inclusive, will be acceptable. (*Note:* The smaller diameter is calculated as 50 mm − 0.25 mm = 49.75 mm, and the larger diameter is calculated as 50 mm + 0.25 mm = 50.25 mm.)

Describing tolerances is made easy and concise with the use of absolute value inequalities. Let the variable A represent the actual dimension of a manufactured part. Also, let the constant t be the target dimension for the part and the constant c be the maximum amount of acceptable variation. Then a manufacturing tolerance for the actual dimension A of a manufactured part can be described by the absolute value inequality $|A - t| \le c$. In the bolt example given above, the variable A is the actual diameter of a bolt coming off the production line, t is the target diameter of 50 mm, and c is 0.25 mm. The manufacturing tolerance for these bolts can be described as $|A - 50| \le 0.25$.

a. Solve the absolute value inequality $|A - 50| \le 0.25$ for A, and write the solution set in interval notation.

b. Explain the meaning of the solution set you found in part *a* in the context of the bolt example described above.

c. Suppose that a manufacturing plant produces large cartons of yogurt, and quality control technicians at the plant make sure that the cartons are neither over filled nor under filled. Each carton is to be filled with a target weight of 32 ounces of yogurt. Manufacturing engineers specify that filled yogurt cartons are considered acceptable if their weights are within 0.4 ounce of the target weight. Write an absolute value inequality that describes the manufacturing tolerance for the weights of the filled yogurt cartons.

d. Solve the absolute value inequality you found in part *c,* and write the solution set in interval notation.

e. Explain the meaning of the solution set you found in part *d* in the context of the yogurt manufacturing situation.

f. A machine shop cuts copper pipe into standard lengths. The target length for the copper pipes is 152.6 cm, and the maximum amount of acceptable variation is 0.15 cm. A quality control technician selects and measures cut pipes at random. If the length of any randomly selected pipe is unacceptable, then the pipe must be rejected. If the technician rejects 3 or more pipes over the course of an 8-hour shift, the production line must be shut down to recalibrate the cutting machines. At the end of an 8-hour shift, a technician has created the following chart.

Sample	Length	Accept?	Reject?
A1	152.51		
A2	152.73		
A3	152.54		
B1	152.72		
B2	152.61		
B3	152.72		
C1	152.69		
C2	152.98		
C3	152.63		
D1	152.59		
D2	152.60		
D3	152.69		

Sample	Length	Accept?	Reject?
E1	152.70		
E2	152.45		
E3	152.69		
F1	152.92		
F2	152.54		
F3	152.63		
G1	152.67		
G2	152.55		
G3	152.79		
H1	152.62		
H2	152.33		
H3	152.48		

Using the data given in the description above, decide whether each pipe sample should be accepted or rejected based on its length. For each sample, place an X in the appropriate column of the chart.

Should the production line be shut down at the end of this shift for recalibration? Explain your reasoning.

4. Group Activity—Analyzing a Municipal Budget

Nearly all cities, towns, and villages operate with an annual budget. Budget items might include expenses for fire and police protection as well as for street maintenance and parks. No matter how big or small the budget, city officials need to know if municipal spending is over or under budget.

Suppose you are an intern hired by Anytown, a small municipality in New England. Each year Anytown creates a municipal budget. The next year's annual budget is submitted for approval by Anytown's citizens at the annual town meeting. In the process of creating next year's budget, you have been tasked with analyzing the current year's budget along with actual spending over the same period. The Anytown town manager wants to take a closer look at any budget categories for which actual spending was drastically different from the amounts that were budgeted. She has asked you to flag (that is, mark or denote) any budget categories in which actual spending differed from the budgeted amount by more than 12% of the budgeted amount.

Below on the left is a copy of the Anytown budget that was approved by citizens for the past year, and on the right is a listing of the actual spending for the same period.

Approved Anytown Budget		Anytown Actual Spending	
Department/Program	**Amount Budgeted**	**Department/Program**	**Actual Spending**
Board of Health		**Board of Health**	
Immunization programs	$25,000	Immunization programs	$23,200
Inspections	$75,000	Inspections	$62,300
Fire Department		**Fire Department**	
Equipment	$670,000	Equipment	$557,200
Salaries	$410,000	Salaries	$398,900
Libraries		**Libraries**	
Media purchases	$135,000	Media purchases	$159,400
Equipment	$45,000	Equipment	$43,100
Salaries	$180,000	Salaries	$175,900
Parks and Recreation		**Parks and Recreation**	
Maintenance	$105,000	Maintenance	$122,600
Playground equipment	$75,000	Playground equipment	$66,900
Salaries	$210,000	Salaries	$175,300
Summer programs	$120,000	Summer programs	$143,000
Police Department		**Police Department**	
Equipment	$450,000	Equipment	$487,400
Salaries	$600,000	Salaries	$601,800
Public Works		**Public Works**	
Recycling	$75,000	Recycling	$71,500
Sewage	$150,000	Sewage	$137,400
Snow removal & salt	$300,000	Snow removal & salt	$398,700
Street maintenance	$375,000	Street maintenance	$422,000
TOTAL	$4,000,000	**TOTAL**	$4,045,400

a. For each category in the budget, write a specific absolute value inequality that describes the condition that must be met for the budget category to be flagged for a closer look by the town manager. In each case, let the variable *x* represent the actual spending for a budget category.

b. For each category in the budget, write an equivalent compound inequality for the condition described in part *a.* Again, let the variable *x* represent the actual spending for a budget category.

c. Using the inequalities from either part *a* or part *b* and the figures from the Anytown Actual Spending table on the first page, complete the following budget worksheet. (The first category has been filled in for you.)

BUDGET WORKSHEET

Budget Category	Budgeted Amount	Minimum Allowed	Actual Spending	Maximum Allowed	Flag?	Amount over/under budget
Immunization programs	$25,000	$22,000	$23,200	$28,000	No	Under $1800

d. Using the Budget Worksheet from part *c,* which budget categories should be flagged for a closer look by the town manager?

e. For each of the flagged budget categories identified in part *d,* explain whether the category was flagged for actual spending being too low or too high and suggest possible reasons why spending in those categories was so far under or over budget.

f. Based on your analysis of the approved Anytown budget for the past year along with the actual spending for the same period, what recommendations would you make for next year's budget? Explain your reasoning.

1. Extension Exercise—The Pendulum

Materials needed: string (at least 1 meter long), weight, meter stick, stopwatch, calculator

In this activity, you will make a simple pendulum by tying a string tightly to a weight and measuring your pendulum's period with a stopwatch. The period of a pendulum is defined as the time it takes the pendulum to complete one full back-and-forth swing. Because the periods will be only a few seconds long, it will be more accurate for you to measure the time for a total of five complete back-and-forth swings and then find the average time for one complete swing. You can change the length of your pendulum by shortening or lengthening the string tied to the weight.

a. For each of the pendulum (string) lengths given in the table below, measure the time required for 5 complete swings and record it in the second column. Next, divide this value by 5 to find the measured period T_m of the pendulum for the given length and record it in the third column of the table.

The theoretical formula relating a pendulum's period T (in seconds) to its length L (in centimeters) is $T = 2\pi\sqrt{\dfrac{L}{980}}$. Use this formula to calculate the theoretical period T for the same pendulum length and record it in the fourth column (rounded to two decimal places). In the last column, find and record the absolute value of the difference between the theoretical period T and the measured period   .

Length L (cm)	Time for 5 Swings (seconds)	Measured Period T_m (seconds)	Theoretical Period T (seconds)	Difference $\lvert T - T_m \rvert$
30				
55				
70				

b. For each period T listed in the table below, use the formula $T = 2\pi\sqrt{\dfrac{L}{980}}$ to calculate the theoretical pendulum length L required to obtain the given pendulum period. Record the value of L (rounded to one decimal place) in the second column.

Next, use the calculated length L for your string pendulum and set it in motion. Measure the time for 5 complete swings and record it in the third column. Divide this value by 5 to find the measured period T_m, and record it in the fourth column. Finally, find and record the absolute value of the difference between the theoretical period T and the measured period T_m.

Period T (seconds)	Theoretical Length L (centimeters)	Time for 5 Swings (seconds)	Measured Period T_m (seconds)	Difference $\lvert T - T_m \rvert$
1				
1.25				
2				

c. Use the general trends that you find in the tables from parts *a* and *b* to describe the relationship between a pendulum's period and its length.

d. Write a paragraph explaining the factors that likely contributed to the differences you found between the values of the theoretical periods and the measured periods.

2. Exploration Activity—Accident Reconstruction

When investigating a motor vehicle accident, law enforcement officials must reconstruct the accident and establish the speed a vehicle was traveling at the time of the accident. If a car left skid marks on the road during the accident, it is possible to approximate the car's speed based on the distance that the car skidded before it came to a stop.

The function $S(x) = \sqrt{30kx}$ is used to estimate the speed S (in miles per hour) that a car was traveling initially if the car stopped after leaving skid marks of length x feet. The value of the constant k depends on the type of pavement and the road conditions. The table below gives values of k for four combinations of road conditions and pavement types.

Road Conditions/ Pavement Type	Value of k
Dry concrete	$k = 0.8$
Wet concrete	$k = 0.35$
Dry asphalt	$k = 1.0$
Wet asphalt	$k = 0.44$

a. Substitute the appropriate value of k into the function $S(x) = \sqrt{30kx}$ and simplify to find a specific expression for speed as a function of x for each combination of road conditions and pavement type listed in the table below.

Road Conditions/ Pavement Type	$S(x) = \sqrt{30kx}$
Dry concrete	$S(x) =$
Wet concrete	$S(x) =$
Dry asphalt	$S(x) =$
Wet asphalt	$S(x) =$

b. Use the functions you found in part *a* to approximate the speed a car was traveling for skid marks of length 150 feet, 200 feet, 250 feet, and 300 feet on dry concrete, wet concrete, dry asphalt, and wet asphalt. Complete the following table with the results, rounding speeds to the nearest mile per hour.

Length *x* of Skid Mark (ft)	Speed *S*(*x*) on Dry Concrete (mph)	Speed *S*(*x*) on Wet Concrete (mph)	Speed *S*(*x*) on Dry Asphalt (mph)	Speed *S*(*x*) on Wet Asphalt (mph)
150				
200				
250				
300				

c. Use a calculator (scientific or graphing) to graph the four versions of the function $S(x)$ that you found in part *a.* Show the graphs on the grid below. Clearly label which curve represents dry concrete, wet concrete, dry asphalt, and wet asphalt.

Speed $S(x)$ (mph)

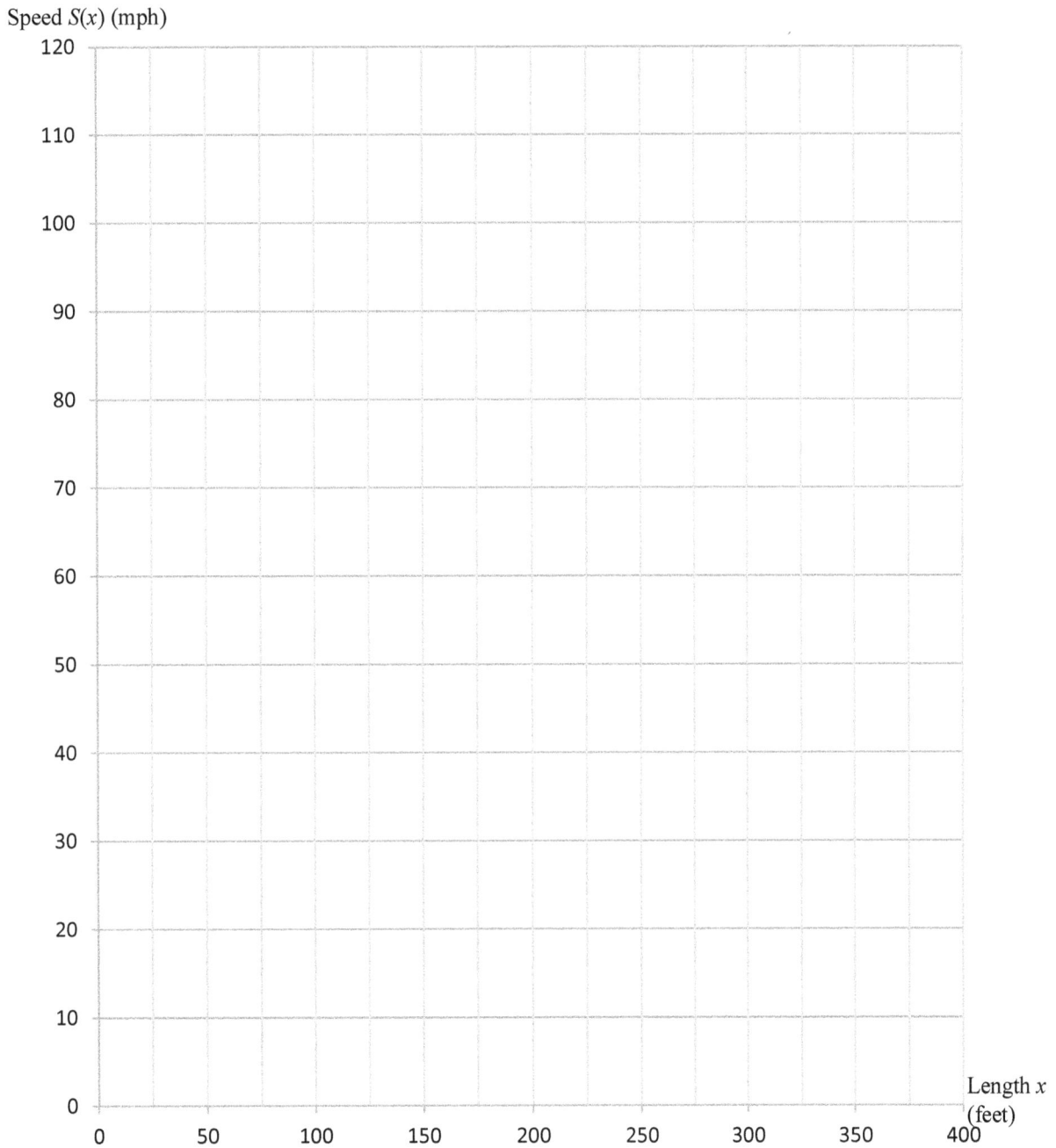

d. Use the graph in part *c* to determine the approximate skid mark length *x* that corresponds to each speed listed in the table below for each set of driving conditions. In motor vehicle accident reconstruction, the skid mark length *x* is equivalent to a vehicle's stopping distance.

Speed (mph)	Stopping Distances			
	Length *x* (ft) on Dry Concrete	Length *x* (ft) on Wet Concrete	Length *x* (ft) on Dry Asphalt	Length *x* (ft) on Wet Asphalt
30				
65				
100				

e. At a particular speed, the safest combination of road conditions and pavement type is the one that yields the shortest stopping distance. From the graphs and data in parts *c* and *d,* which combination appears to be the safest and which appears to be the most dangerous? Explain your reasoning.

f. What other factors might contribute to the length of skid marks left by a car when it is stopped suddenly?

3. Conceptual Exercise—Diffusion

Diffusion is the spontaneous movement of the molecules of a substance from a region of higher concentration to a region of lower concentration until a uniform concentration throughout the region is reached. For example, if a drop of food coloring is added to a glass of water, the molecules of the coloring are diffused so that eventually the entire glass of water is colored evenly without any kind of stirring. (Of course, stirring speeds up this process.) Diffusion is also mostly responsible for the spread of the smell of baking brownies throughout a house.

Diffusion is used or seen in important aspects of many disciplines. The following list describes situations in which diffusion plays a role.

- In the commercial production of sugar, sugar can be extracted from sugar cane through a diffusion process.
- Solid-state diffusion plays a role in the manufacturing process of silicon computer chips.
- In biology, the diffusion phenomenon allows water molecules, nutrient molecules, and dissolved gas molecules (such as oxygen and carbon dioxide) to pass through the semipermeable membranes of cell walls.

a. In chemistry, Graham's law states that the diffusion rate of a substance in its gaseous state is inversely proportional to the square root of its molecular weight. Write an equation for the relationship described by Graham's law. Be sure to define the variables and constants that you use.

b. According to Graham's law, which molecule will diffuse more rapidly: molecule A with a molecular weight of 58.4 or molecule B with a molecular weight of 180.2? Explain your reasoning.

c. Another useful property of diffusion is that the distance a material diffuses over time is directly proportional to the square root of the time. Write an equation for the relationship between the distance that a material diffuses and time. Again, be sure to define the variables and constants that you use.

d. Suppose it takes sugar 1 week to diffuse a distance of 1 cm from its starting point in a particular liquid. How long will it take the sugar to diffuse a total of 3 cm from its starting point in the liquid?

e. (Optional) Research at least two other situations in which diffusion plays a role. Explain any formulas or equations associated with the situations.

4. Group Activity—Maximum Volume of a Cone

<u>Materials needed</u>: paper, compass, string, scissors, tape, ruler

In this activity, you will determine what size wedge (shaped like a slice of pie) to cut from a paper circle to form what is left of the circle into a cone of the largest possible volume.

a. Use a compass to draw four circles, each with a 5-inch radius, on sheets of paper. Be sure to mark the center of each circle. Cut out each circle with scissors. Compute the circumference c, rounded to the nearest tenth, of one of these identical circles. *Hint:* Use the formula $c = 2\pi r$ to compute the circumference of a circle from its radius r.

Circumference of a circle with a 5-inch radius: $c \approx$ _____ inches

b. Cut a piece of string that is $x = 4$ inches long. On one paper circle, mark a 4-inch long curved segment along the circumference of the circle by laying the string along the circle's edge. Draw a wedge from the center of the circle to the ends of the string. Cut out the resulting wedge. Use tape to form what is left of the circle into a cone without overlapping the edges.

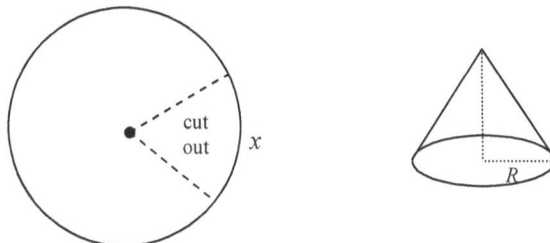

Repeat this process with the remaining circles using string lengths of $x = 6$ inches, $x = 8$ inches, and $x = 10$ inches.

Because the base of each resulting cone is a circle, the radius R and circumference C of the cone's base are related by the formula $2\pi R = C$. Solve this equation for R in terms of C. [*Note:* capital letters are used here to distinguish the circumference and radius of the base of the cone from those of the original circles.]

$R =$ _____

c. For each of the four cones created in part *b,* compute the circumference *C* and radius *R* of the base of the cone. The circumference *C* of the base of the cone is given by the difference between the circumference *c* of the original circle (found in part *a*) and the length *x* of the curved edge that was cut from the circle. Use the equation found in part *b* to compute the radius *R* of the base of the cone (to the nearest tenth). Record your results in the table.

Length *x* of Curved Edge of Wedge (inches)	Circumference *C* of Base of Cone $C = c - x$ (inches)	Radius *R* of Base of Cone (inches)
4		
6		
8		
10		

d. A vertical cross section through the vertex of each cone includes the right triangle shown below with base *R* and height *h.* The height *h* of this triangle coincides with the height *h* of the cone.

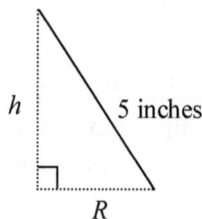

Use the Pythagorean theorem to write and solve an equation for *h.*

e. The volume of a cone is given by the formula $V = \frac{1}{3}\pi R^2 h$. Substitute the expression for h (found in part *d*) into this volume formula to obtain a formula that is in terms of only R.

f. Copy the values of R that you found in part *c* into the second column of the table below. Then use the formula that you found in part *e* to compute the volume (rounded to the nearest tenth) of each of the four cones and record the results in the third column of the table.

Length *x* of Curved Edge of Wedge (inches)	Radius *R* of Base of Cone (inches)	Volume of Cone (cubic inches)
4		
6		
8		
10		

g. For the four cones that your group made, what was the largest volume? Which value of *x* was associated with the largest volume?

Largest volume = _____

x = _____

h. If the cone of largest volume were formed from a circle with a radius of 5 inches, would the wedge removed from the circle to make the cone have one of the values for *x* (4, 6, 8, or 10) that you used in this activity? Or would the value of *x* leading to the largest volume lie somewhere between these values? Discuss this question with your group and summarize your conclusions below.

1. Extension Exercise—Ball Drop

April is standing on the edge of a cliff that overhangs a river. She wants to know how high the cliff is above the water's surface, so she drops a tennis ball and times how long it takes for the ball to hit the water. Four seconds after dropping the ball, she sees it hit the water. April knows that the ball's height (in feet) above the river t seconds after it is released can be calculated by the formula

$$s(t) = -16t^2 + s_0,$$

where s_0 is the height above the river from which the ball was dropped.

April's first task is to determine the value of s_0, the height of the cliff above the river.

a. Because $s(t)$ is the height of the ball above the river after t seconds, use the given information and the meaning of _____ to determine the values of _____ and t when the ball first hits the water. Explain how you determined these values.

          _____ = _____ feet $t =$ _____ seconds

b. Substitute the values from part *a* for _____ and t into the equation $s(t) = -16t^2 + s_0$, and solve the resulting equation for s_0 to find the height of the cliff above the river.

          Equation after substitutions: _____

          Solution: s_0 = _____ feet above the river.

c. Substitute the value for s_0 from part *b* into the formula $s(t) = -16t^2 + s_0$. Write the resulting function below.

          _____ = _____

d. Graph the parabola defined by $s(t)$.

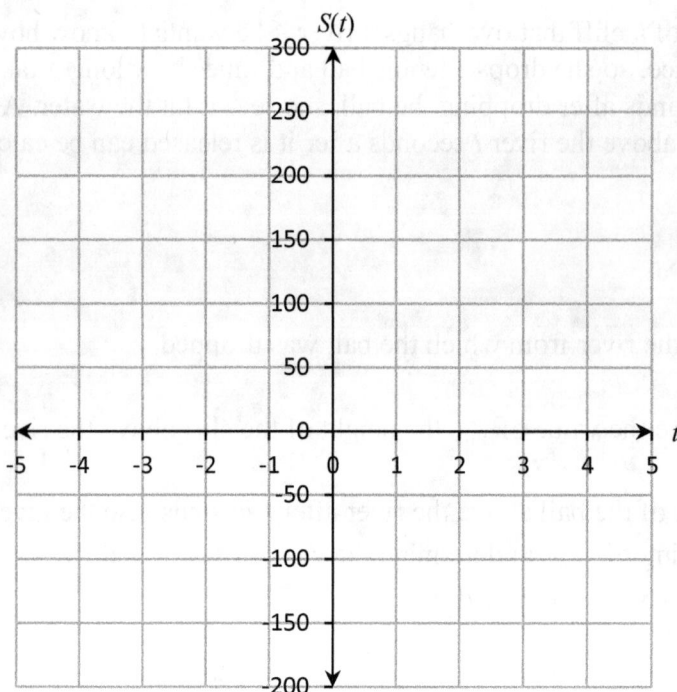

e. For each of the following situations, set up and solve an appropriate equation to find out how long it takes the ball to reach each of the given heights above the water and how far the ball has fallen when it reaches that height. Write an ordered pair for each situation and plot it on the graph above. Disregard the negative solutions. Round results to the nearest tenth.

40 feet above the water:

equation: _____

time of fall: _____

ordered pair: _____

distance fallen: _____

Halfway down the cliff:

equation: _____

time of fall: _____

ordered pair: _____

distance fallen: _____

f. Given the physical constraints of the function $s(t)$ in the context of the falling ball, use part *d* to find the domain and range of the function that describes the falling ball. Explain your answer.

2. Exploration Activity—Recognizing Linear and Quadratic Models

We have seen throughout the text far that data can be modeled by both linear models and quadratic models. However, when we are given a set of data to model, how can we tell which type of model—linear or quadratic—is appropriate? The best choice depends on looking at a graph of the data. If the plotted points fall roughly on a straight line, then a linear model is usually the better choice. If the plotted data points seem to fall on a definite curve or if a maximum or minimum point is apparent, a quadratic model is usually the better choice.

Two different sets of data are given in the tables below. One of the sets of data shown in these tables is best modeled by a linear function and one is best modeled by a quadratic function. **Table A** summarizes the box office receipts y (in dollars) that a community theater group has earned at recent performances when it set its ticket price x (in dollars) at various levels. For instance, the table shows that for a recent performance at which the group charged $15 per ticket, a total of $121,000 was collected at the box office in ticket sales. **Table B** summarizes the relationship between the average monthly rent y (in dollars) charged for an apartment in a major U.S. city and the size of the apartment x (in square feet). For instance, in this particular city the average monthly rent of an apartment measuring 1000 square feet is $3240.

Table A							
Ticket price, x	10	15	20	25	30	35	40
Box office receipts, y	80,000	121,000	148,000	172,000	180,000	180,000	168,000

Table B								
Square footage, x	900	950	1000	1050	1100	1150	1200	1250	1300
Monthly rent, y	2925	3040	3240	3444	3630	3740	3840	4030	4225

a. Plot the data from **Table A** as ordered pairs on the grid below.

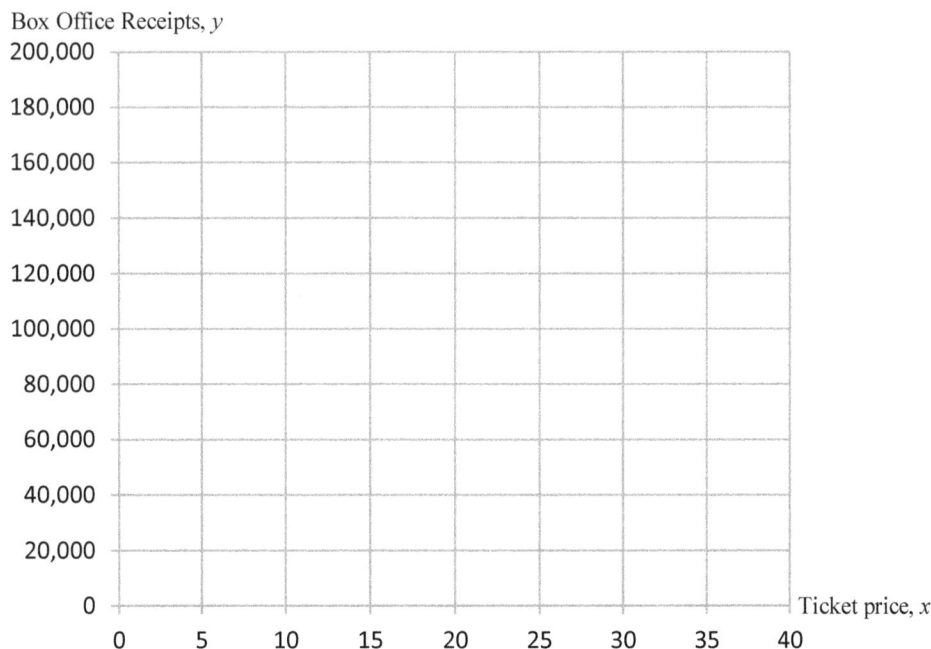

Box Office Receipts, y

b. Plot the data from **Table B** as ordered pairs on the grid below.

Monthly rent, *y*

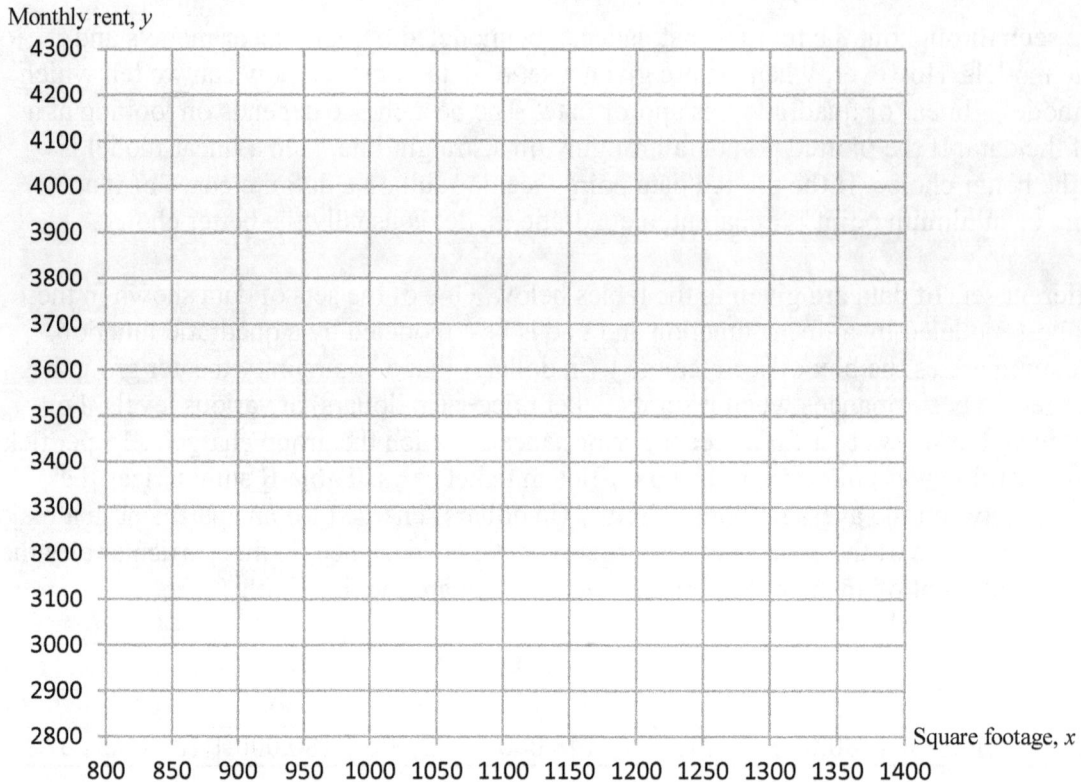

Square footage, *x*

c. Based on your graphs in parts *a* and *b*, which type of model, linear or quadratic, do you think should be used for each set of data? Explain your reasoning.

d. For the set of data that you have determined to be linear, find a linear function that fits the data points. Explain the method you used. (*Hint:* You might begin by drawing a line that seems to best "fit" the points, then use the methods of Chapter 10 to find the equation of your line.)

e. For the set of data that you have determined to be quadratic, estimate the location on your graph of the data where you believe the vertex of the parabola would occur. Substitute the coordinates of this vertex into the quadratic model $f(x) = a(x-h)^2 + k$.

f. From part *e*, solve for the remaining unknown constant *a* in the quadratic model by substituting the coordinates for another data point into the function. (If needed, round the value of *a* to one decimal place.) Write the final form of the quadratic model for this data set in the form $f(x) = a(x-h)^2 + k$, as well as in standard form, $f(x) = ax^2 + bx + c$.

g. Use your model for **Table A** to estimate box office receipts if ticket price is set at $23.

h. Use your model for **Table B** to estimate the monthly rent for an apartment with a size of 1080 square feet.

i. For the set of data that you have determined to be linear, enter the data into a graphing calculator and use the linear regression feature (consult the user's manual for your graphing calculator for details) to find the best-fit linear function that models the data. Compare this function with the linear function that you found by hand. How are they alike or different?

j. For the set of data that you have determined to be quadratic, enter the data into a graphing calculator and use the quadratic regression feature (consult the user's manual for your graphing calculator for details) to find the best-fit quadratic function that models the data. Compare this function with the quadratic function that you found by hand. How are they alike or different?

3. Conceptual Exercise—The World's Largest Non-Steerable Radio Telescope

Because a parabola has nice reflecting properties, the shape of a parabola is used in many kinds of telescopes. The world's largest non-steerable radio telescope is Arecibo Observatory in Puerto Rico. This telescope consists of a huge parabolic dish built into a valley. The cross-section of Arecibo's parabolic shape through the center of the dish is described by the function $f(x) = 0.000668x^2 - 167$, where both x and $f(x)$ represent distances in feet. A graph of this equation is shown below.

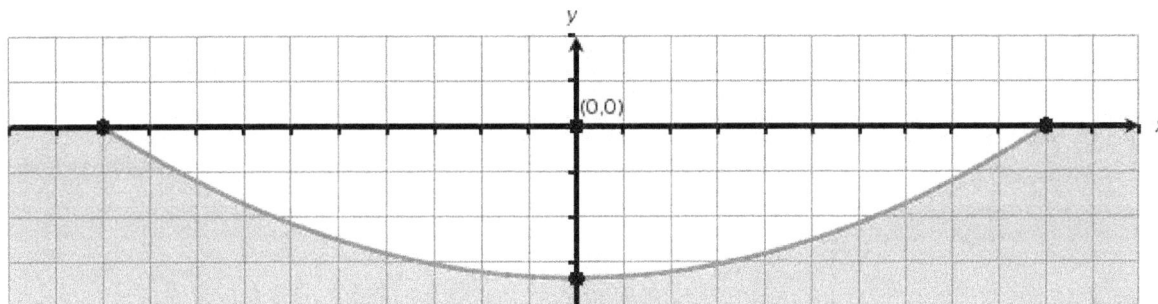

a. Find the x-intercepts of the parabola and label them on the graph.

b. How wide is Arecibo's parabolic dish at its top? Explain your reasoning.

c. Find the vertex of the parabola and label it on the graph.

d. What is the maximum depth of Arecibo's parabolic dish? Explain your reasoning.

e. What is the depth of Arecibo's parabolic dish at a distance of 150 feet from its center?

f. At what distance from the center of Arecibo's parabolic dish is its depth 100 feet? Round your answer to the nearest tenth of a foot.

4. Group Activity—Graphs of Quadratic Functions

In this activity you will use a graphing calculator to investigate the graphs of quadratic functions. Keep in mind the following properties as you work through this activity:

- A quadratic function has the form $f(x) = ax^2 + bx + c$ where $a \neq 0$.
- If $a > 0$, the graph (called a parabola) opens upward. If $a < 0$, the parabola opens downward.
- The vertex of the parabola is the highest point (maximum) if the parabola opens downward, or is the lowest point (minimum) if the parabola opens upward. The coordinates of the vertex are $\left(\dfrac{-b}{2a}, f\left(\dfrac{-b}{2a} \right) \right)$.

a. Complete the table given below. For each quadratic function in the first column, record the value of *a,* and determine whether the related parabola opens upward or downward. Then calculate the coordinates of the vertex (rounded to the nearest tenth). Finally, let each function $f(x)$ equal 0 and solve for *x*. These solutions x_1 and x_2 are also called **roots** of the function. Round real roots to the nearest tenth. Leave complex roots in complex form.

$f(x)$	Value of a	Opens up or down?	Vertex Coordinates	Roots x_1 and x_2
$f(x) = 5x^2 - 3x - 9$				
$f(x) = -15x^2 - 11$				
$f(x) = 14x^2 + 12x + 5$				
$f(x) = -8x^2 + 3x + 7$				

b. Which of the functions in the table have real roots and which have complex roots?

Functions with real roots: _____

Functions with complex roots: _____

c. On a graphing calculator, graph each of the functions in the table. Use a viewing window which clearly shows both the shape of the parabola and the *x*-axis. Use the trace feature to find the vertex and *x*-intercepts (if there are any) of each parabola. Record your results below, rounded to the nearest tenth. Make a rough sketch of each graph, labeling the vertex and the *x*-intercepts if there are any.

$f(x) = 5x^2 - 3x - 9$

    vertex: (_____, _____)

    intercepts (if any): (_____, _____)   (_____, _____)

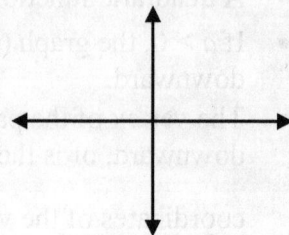

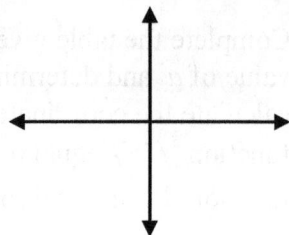

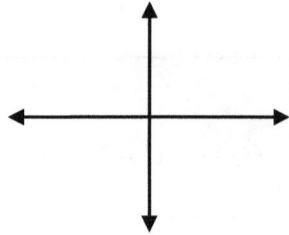

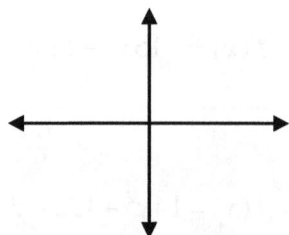

$f(x) = -15x^2 - 11$

    vertex: (_____, _____)

    intercepts (if any): (_____, _____)   (_____, _____)

$f(x) = 14x^2 + 12x + 5$

    vertex: (_____, _____)

    intercepts (if any): (_____, _____)   (_____, _____)

$f(x) = -8x^2 + 3x + 7$

    vertex: (_____, _____)

    intercepts (if any): (_____, _____)   (_____, _____)

d. Explain the connection between the real number roots of a quadratic function and the *x*-intercepts of its graph. When a quadratic function has complex roots, what can be said about the *x*-intercepts of its graph?

Suppose $a > 0$ in $f(x) = ax^2 + bx + c$, and the vertex of the parabola is $(-17, 4)$. In parts *e–h* below, determine if the statement is true, false, or whether not enough information is available to make a decision. Show or explain how you decided.

e. The equation $0 = ax^2 + bx + c$ has two real solutions.

    Circle one: True   False   Not enough information

f. The point $(-17, 4)$ is the highest point on the graph of   .

    Circle one: True   False   Not enough information

g. $\dfrac{-b}{2a} = -17$

    Circle one: True   False   Not enough information

h. $f(5) = 22$

    Circle one: True   False   Not enough information

The graph of $f(x) = ax^2 + bx + c$ has a vertex in quadrant II and opens downward. In parts *i–k* below, determine if the statement is true, false, or whether not enough information is available to make a decision. Show or explain how you decided.

i.   The equation $0 = ax^2 + bx + c$ has two real solutions. One is negative, and the other positive.

      Circle one: True   False   Not enough information

j.   For the function $f(x) = ax^2 + bx + c, \; a > 0$.

      Circle one: True   False   Not enough information

k.   The vertex of $f(x) = ax^2 + bx + c$ is $(-2, 5)$.

      Circle one: True   False   Not enough information

1. Extension Exercise—Growth Limits

The asexual reproduction of cells, known as *mitosis*, is a process that allows a cell to divide into two cells such that each new cell has DNA that is identical to that of the original cell. This process allows the genetic information contained in a single cell to be copied again and again.

a. Suppose that you have set up an experiment that places 100 cells under lab conditions allowing each to divide into two cells every 20 minutes ($\frac{1}{3}$ of an hour). Complete the table below to see how cells in this experiment multiply over time.

Time t (hours)	Cell Population
0	100
$\frac{1}{3}$	
$\frac{2}{3}$	
1	
$1\frac{1}{3}$	

b. For the experiment described in part *a*, use the general exponential model given below to find a function that represents the growth of the cell population P with respect to time t. Note that P_0 is the initial population at time $t = 0$, and $\dfrac{1}{k}$ is the time that it takes for the population to double.

$$P(t) = P_0 \cdot 2^{kt}$$

c. How many cells are in the population after 3 hours?

d. Suppose there are 409,600 cells when you leave for lunch. Upon returning you find that there are 3,276,800 cells. How long were you gone?

e. Why might it be problematic to construct a graph that shows the population size over an 8-hour period?

f. Explain why an exponential growth model of any population—whether it is cells, rabbits, or humans—can only make accurate predictions within a restricted time interval.

g. Suppose that an initial population of 1000 cells grows to 16,000 cells in two hours. What is the doubling time of this population? Use $P(t) = P_0 \cdot 2^{kt}$ to find k. Then recall that $\dfrac{1}{k}$ is the doubling time, or the time that it takes the population to double.

h. Suppose an initial population of 1000 cells grows to 17,000 cells in two hours. What is the doubling time of this population? Round your answer to two decimal places.

2. Exploration Activity—Modeling Population Growth

The table below gives the total midyear population of Egypt from 2001 – 2007 rounded to the nearest thousand. This activity will explore how exactly this population is increasing.

Time t (year)	Population P	Annual Change in Population	Growth Factor
0 (2001)	66,852,000		
1 (2002)	68,265,000		
2 (2003)	69,734,000		
3 (2004)	71,211,000		
4 (2005)	72,688,000		
5 (2006)	74,207,000		
6 (2007)	75,814,000		

(*Source:* U.S. Census Bureau, International Data Base)

a. To get a clear idea of how the population is increasing, find the annual change in population between consecutive years. Write each result in the third column, and then complete the sentence below with one of the following words: *decreasing, constant,* or *increasing.*

The population is increasing at a(n) _____ rate.

b. If a population is growing exponentially over a period of time, then its growth factor should be (relatively) constant from year to year. Calculate Egypt's growth factors by dividing each year's population by the previous year's population. Write the result in the fourth column of the table. Round your results to the thousandths place.

$$\text{Growth Factor} = \frac{\text{Current Year Population}}{\text{Previous Year Population}}$$

What can you conclude about how the population of Egypt was growing during these years based on the growth factors you calculated?

c. A population's growth factor over a period of time is given by $(1+r)$, where r is the population's **growth rate,** or constant percent of increase it experienced during that period. What would you say was Egypt's annual growth rate r from 2001 to 2007? (Express r as a percent.) Explain your reasoning.

d. In general, we can express the exponential growth of a population P as a function of time t using the model $P(t) = P_0 \cdot (b)^t$, where P_0 is the initial population at $t = 0$, and the base b is the constant growth factor (found in part b). Substitute the appropriate values for P_0 and b into the function $P(t)$ to obtain a model for Egypt's population as a function of time.

$P(t) = $_____, where $t = 0$ corresponds to the year _____

e. Use the model developed in part d to predict Egypt's population in the years 2013 and 2018. Use function notation and show all necessary calculations. Round your results to the nearest thousand.

f. According to the U.S. Census Bureau, the actual population of Egypt (rounded to the nearest thousand) was 87,679,000 in 2013 and 99,413,000 in 2018. Compare these values to your model's predictions from part e and explain what probably happened to Egypt's annual growth rate in the years following 2007.

g. According to the population growth model developed in part d, how long will it take for Egypt's population to become double its 2001 level? *Hint:* Set your model from part d equal to twice the 2001 population and then solve for t by using logarithms. Give an exact answer and then approximate to the nearest whole year.

3. Conceptual Exercise—Acid vs. Alkaline

A quantity known as pH is used as a measure of the acidity or alkalinity of a solution. The pH value is calculated as the negative logarithm of the solution's hydrogen-ion concentration, $[H^+]$, measured in moles per liter. The formula for pH is shown below.

$$pH = -\log([H^+])$$

Swimming pools and aquariums are good examples of water that must be regularly tested to maintain the proper pH level for a healthy environment. Both are very sensitive to small changes in   with pools needing to keep their pH at a level that stops microorganisms from growing, and aquariums needing a pH level that allows microorganisms to grow.

a. What is the pH of vinegar with a hydrogen-ion concentration of $[H^+] = 1 \times 10^{-3}$?

b. What is the pH of orange juice with a hydrogen-ion concentration of $[H^+] = 2 \times 10^{-4}$? Round to the tenths place.

c. This table gives the hydrogen-ion concentration   (in moles per liter) of a number of solutions. Find the pH of each solution, and record the results in the third column.

Solution	(mol/L)	pH
Gastric Juice	1×10^{-1}	
Black Coffee	1×10^{-5}	
Pure Water	1×10^{-7}	
Household Ammonia	1×10^{-11}	

d. The figure below gives the pH scale of acidity and alkalinity. Find the value for   that corresponds to each pH and express it as a power of 10 below each vertical mark. Note that the greater the   , the more acidic the solution.

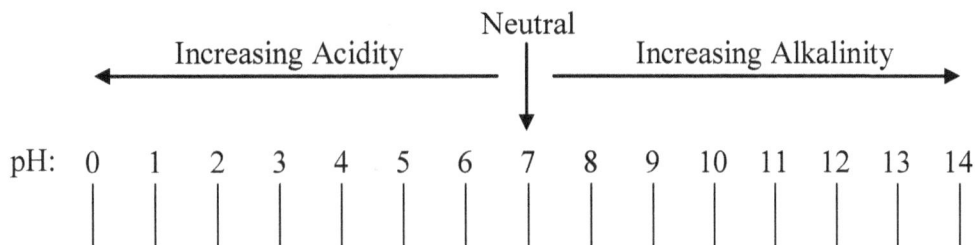

Neutral

← Increasing Acidity — | — Increasing Alkalinity →

pH: 0 1 2 3 4 5 6 7 8 9 10 11 12 13 14

e. Why do you think that pH is measured as the negative logarithm of $[H^+]$ instead of just the hydrogen-ion concentration $[H^+]$ by itself?

f. Complete the following sentence referring to the scale in part *d* as needed.

If pH decreases by one unit, then the hydrogen-ion concentration $[H^+]$_____

_____

g. The pH of tomato juice is 4.1. Set up and solve a logarithmic equation to find the hydrogen-ion concentration $[H^+]$ of tomato juice.

h. Find the hydrogen-ion concentration $[H^+]$ for the solutions in the table below.

Solution	pH	$[H^+]$ (mol/L)
Lemon Juice	2.3	
Beer	4.2	
Milk	6.6	
Blood	7.4	

i. Acid rain in some areas of the northeast has a pH of 4.2 while unpolluted rain has a pH of about 5.6. Find the ratio of the hydrogen-ion concentration in acid rain to that of unpolluted rain.

4. Group Activity—Choosing a Model

Suppose your group is responsible for website administration at a small but growing company. One of your duties is to ensure that your company's newly established website can adequately handle the number of visitors to it. You decide to collect recent website usage statistics and find a mathematical model for the data to help predict future numbers of visitors. The monthly website usage data is shown in the table below. Ultimately, you would like to use the mathematical model based on this data to predict when your website's server capacity must be expanded.

Monthly Website Usage					
Month, x	1	2	3	4	5
Number of visitors, y	166,511	1,320,978	1,996,298	2,475,445	2,847,100

a. The first step in finding an appropriate model for the website usage data is to decide what type of mathematical function to use: linear, quadratic, cubic, exponential, or logarithmic. A plot of the data can be useful in making this choice. Begin by using a graphing calculator to plot the monthly website usage data as ordered pairs using the following viewing window.

Xmin = 0   Ymin = 0
Xmax = 6   Ymax = 3,000,000
Xscl = 1   Yscl= 500,000

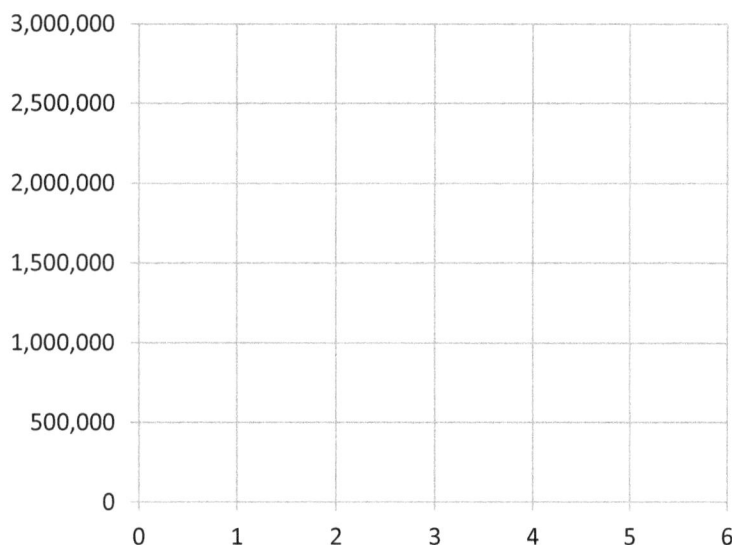

b. Discuss the pattern formed by the plotted points in your graph from part *a*. Which of the function types under consideration do you think might fit the "shape" of this data?

c. Use a graphing calculator to find the best-fit *linear* regression model for the monthly website usage data. Using the viewing window given in part *a*, graph this linear function along with the data points. Comment on how well the model fits the data points.

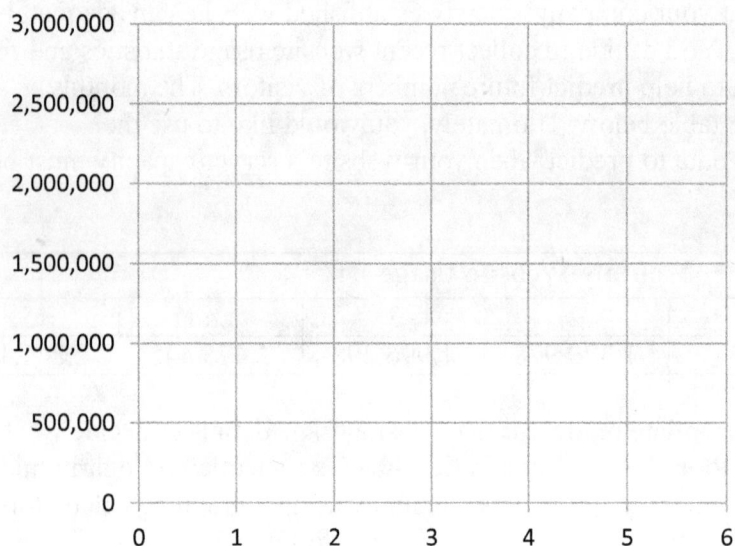

d. Use a graphing calculator to find the best-fit *quadratic* regression model for the monthly website usage data. Using the viewing window given in part *a*, graph this quadratic function along with the data points. Comment on how well the model fits the data points.

e. Use a graphing calculator to find the best-fit *cubic* regression model for the monthly website usage data. Using the viewing window given in part *a*, graph this cubic function along with the data points. Comment on how well the model fits the data points.

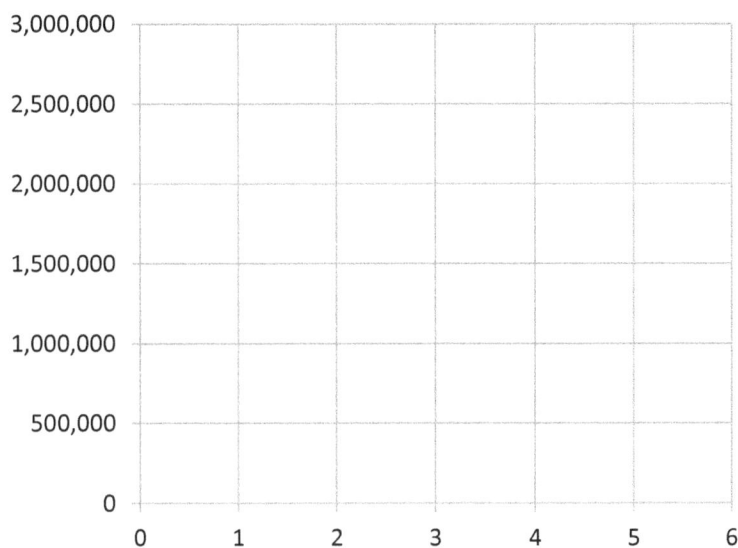

f. Use a graphing calculator to find the best-fit *exponential* regression model for the monthly website usage data. Using the viewing window given in part *a*, graph this exponential function along with the data points. Comment on how well the model fits the data points.

g. Use a graphing calculator to find the best-fit *logarithmic* regression model for the monthly website usage data. Using the viewing window given in part *a*, graph this logarithmic function along with the data points. Comment on how well the model fits the data points.

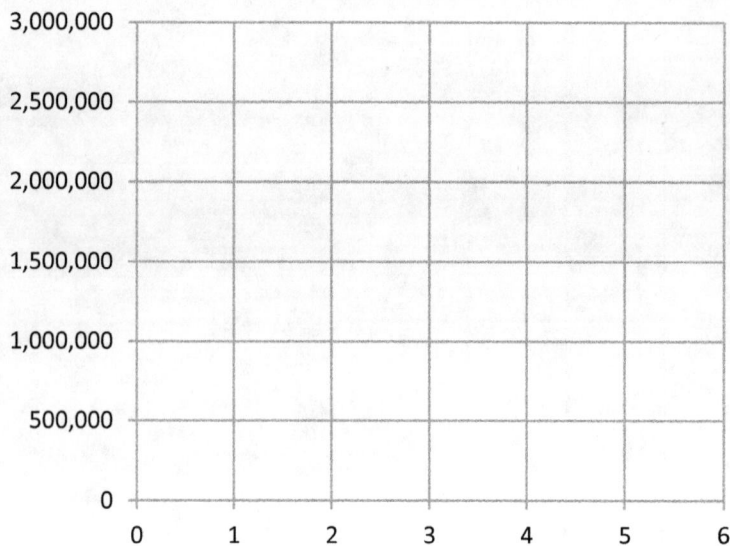

h. Look back at the graphs of the five models you fit to the monthly website usage data in parts *c–g*. As you search for the most appropriate model for the data, are there any models you would choose to eliminate at this point? If so, which one(s) and why?

i. Remember that you are hoping to use your mathematical model to help make some predictions about website usage going into the future. Change your graphing calculator's viewing window to the following settings,

Xmin = 0	Ymin = 0
Xmax = 12	Ymax = 5,000,000
Xscl = 1	Yscl= 500,000

then graph the models remaining under consideration along with the data points. Does this new viewing window change your perspective at all on the appropriateness of the remaining models? Why or why not?

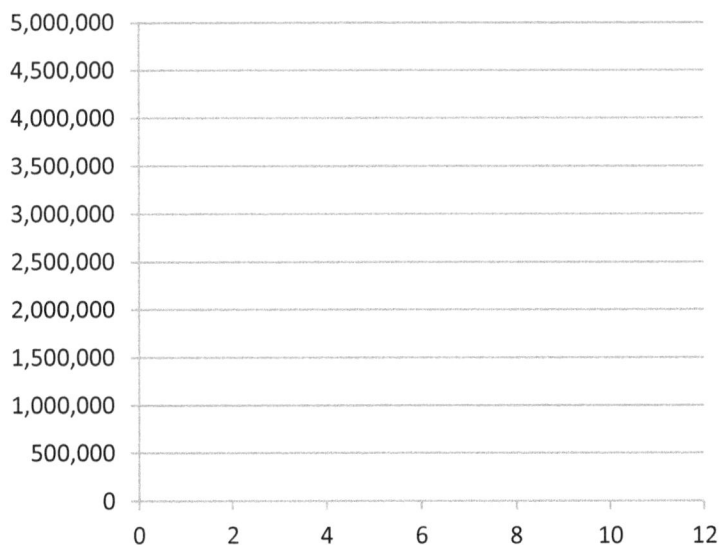

j. Now suppose that some additional time goes by. You find that in month 7 the number of website visitors is 3,407,500. Add this data point to your graphing calculator graph from part *i*. Does this additional data help you select an appropriate model for the monthly website usage data? Why or why not? If you can select only one model, which would you choose? Why?

k. You have determined that your company's website server capacity must be expanded once monthly traffic reaches 5,000,000 visitors per month. Use the model you selected in part *j* to determine after how many months server capacity will need to be expanded. Round your answer to the nearest whole month.

1. Extension Exercise—Sputnik's Orbit

Just as planets move in elliptical orbits with the sun at one focus, satellites can move in elliptical orbits with the center of a planet at one focus. In 1957 the Soviet Union sent the first artificial satellite, Sputnik I, into orbit around Earth. Sputnik's orbit ranged from 132 miles to 583 miles above Earth's surface.

The figure below shows Sputnik's orbit with Earth's center as one of the foci so that $a + c$ is Sputnik's maximum distance from Earth's center, and $a - c$ is Sputnik's minimum distance from Earth's center. This exercise explores how to use this information to find an equation that models Sputnik's orbit around Earth. *Note:* Earth's radius is 3950 miles.

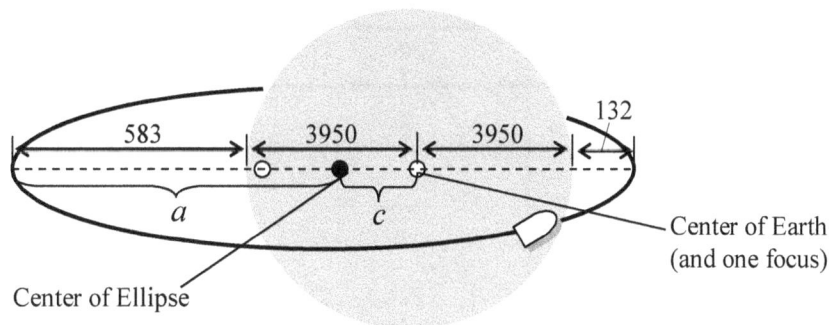

a. If the maximum distance from Sputnik to Earth's center is M and the minimum distance is m, then we have the following system of equations:

$$a + c = M$$
$$a - c = m$$

Substitute values for M and m, then solve the system to find a and c.

b. Use $b^2 = a^2 - c^2$ to find b^2 and b. Round b to the nearest tenth.

c. We can see the nearly circular Earth as if we are looking down from deep space on Sputnik's orbit around Earth by graphing a circle with the origin as its center and a radius of 3950 miles. Graph $x^2 + y^2 = (3950)^2$ on the blank grid below.

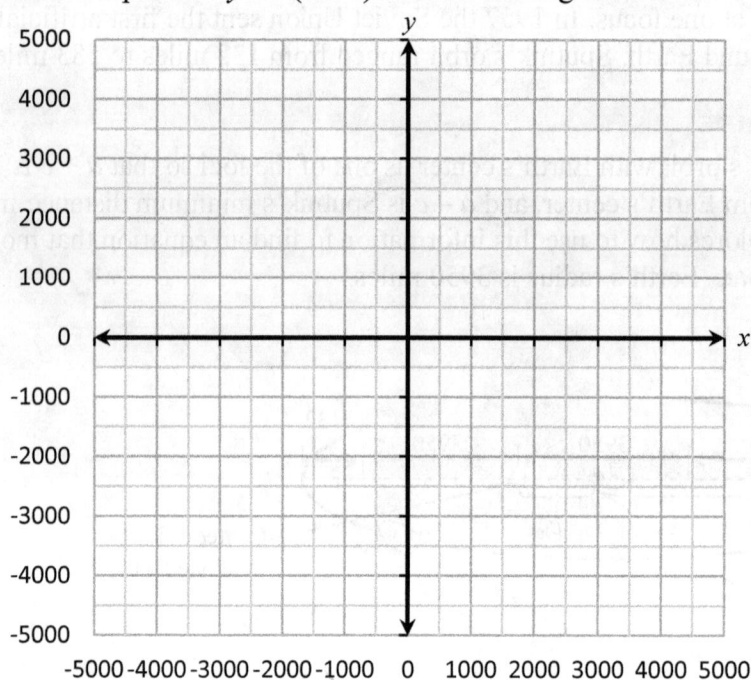

d. Sputnik's orbit can be graphed as an ellipse with the center of Earth at one focus. Find an equation for Sputnik's orbit using the standard form

$$\frac{(x-h)^2}{a^2} + \frac{(y-k)^2}{b^2} = 1.$$

Assume that the x-axis is the orbit's major axis. Also assume that Earth's center is located at the origin and coincides with the right-hand focus of the ellipse. Sputnik's elliptical orbit is centered at (h, k), located on the x-axis at a distance of c units to the left of the origin.

e. Graph the equation for Sputnik's orbit around Earth on the xy-plane in part c.

f. Use the formula $e = \dfrac{c}{a}$ to find the eccentricity e of Sputnik's orbit. Round to the ten-thousandths place.

2. Exploration Activity—Bridge Design

Suppose you are an architect. You have just designed a bridge over a two-lane road. Spanning the road is an arch in the shape of a half-ellipse that is 40 feet wide at the base of the arch and is 15 feet tall at the center of the arch. A colleague has just pointed out that the bridge must have a 13-foot clearance for vehicles on the road. The road is 22 feet wide, and its center line falls directly beneath the highest point of the arch. The following steps will help you decide if you have a problem with your bridge.

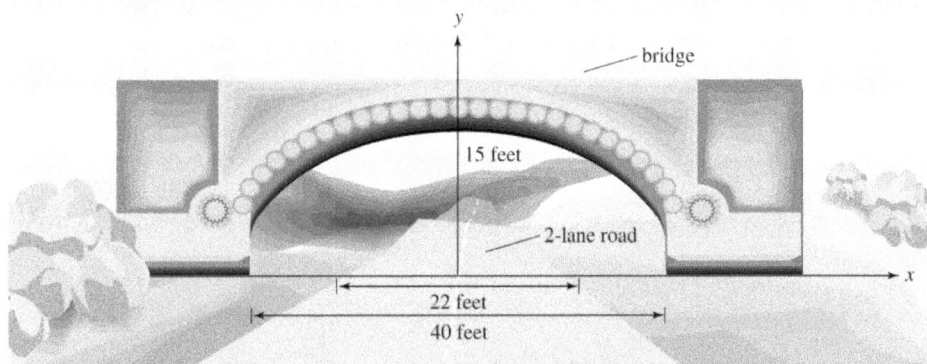

a. Use the information given about the bridge to write an equation for the ellipse that defines the arch.

b. Write an inequality that defines the set of ordered pairs in the coordinate plane that could freely pass beneath the bridge.

c. Envision a 13-foot-tall semi truck passing under the bridge in the right-hand lane of the road. Remember that the truck can drive within any portion of the right-hand land, including along the extreme far right edge of the road. What ordered pair would describe the highest point on a 13-foot tall semi truck that is the farthest to the right on the two-lane road?

3. Conceptual Exercise—Parabolic Reflector

A parabola can be defined as the set of all points that have an equal distance from a fixed point called the focus and a fixed line called the **directrix**. Inside the head of a flashlight you will find a parabolic mirror which reflects the light shining from a bulb at the focus. The light coming from the focus strikes the parabola's edges and reflects out in lines that are parallel to the axis of symmetry. This is demonstrated in the figures below.

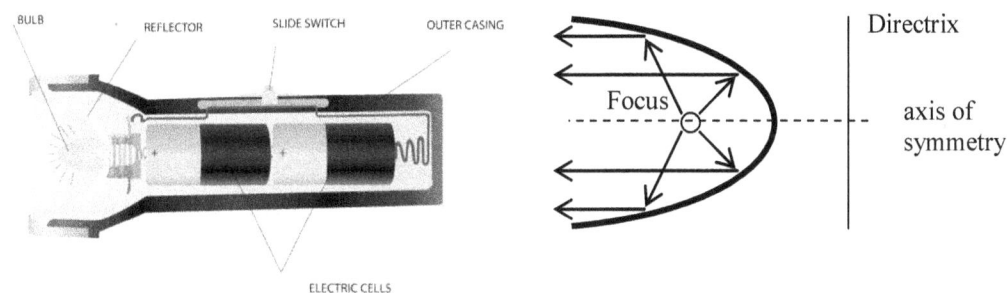

a. Suppose the mirror inside the flashlight is a parabolic solid with a diameter of 4 inches and a depth of 2 inches. The cross section is a parabola that can be modeled by the equation $x^2 = 2y$ on the interval $[-2, 2]$. Sketch this model below.

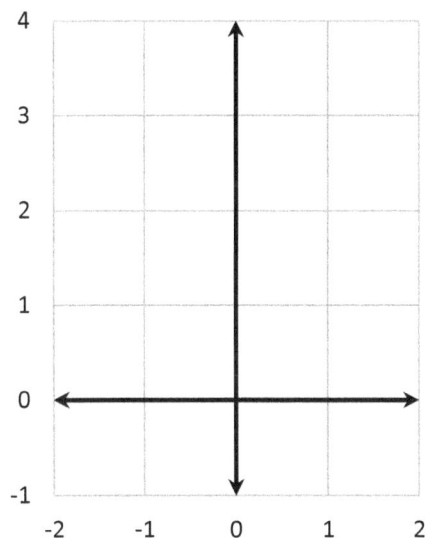

b. Parabolas in the standard form $x^2 = 4py$ have focus $(0, p)$ and directrix $y = -p$. Find the focus and directrix for the equation you graphed in part *a*, then sketch them on your graph.

c. Use the graph in part *a* to illustrate a light bulb at the focus emanating rays that bounce off the parabola's edge and reflect out of the parabola parallel to the axis of symmetry. What is the equation of the line that represents the axis of symmetry?

d. If you keep the diameter of the parabolic mirror at 4 inches but change the depth to 1 inch, then you can model a cross section with the equation $x^2 = 4y$. Graph this equation below, and label the point (focus) where the light bulb must be located in order for the flashlight to reflect light rays that are parallel to the axis of symmetry.

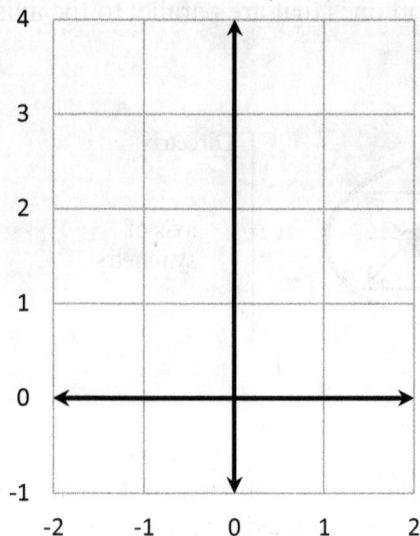

e. If you keep the diameter of the parabolic mirror at 4 inches but change the depth to 0.5 inch, then you can model a cross section with the equation $x^2 = 8y$. Graph this equation below, and label the point (focus) where the light bulb must be located in order for the flashlight to reflect light rays that are parallel to the axis of symmetry.

f. As you decrease the depth of the parabola, describe what happens to the location of the light bulb.

4. Group Activity—Fishing Area

Jessica likes to row her boat out to the middle of Crystal Lake to try her luck at fishing. She has been contemplating buying a new rod and reel that will allow her to make longer casts in any direction and thus increase the area she has for fishing. On the graph below, suppose that her anchored boat is at the origin, and the vertical and horizontal scale measures distance in meters.

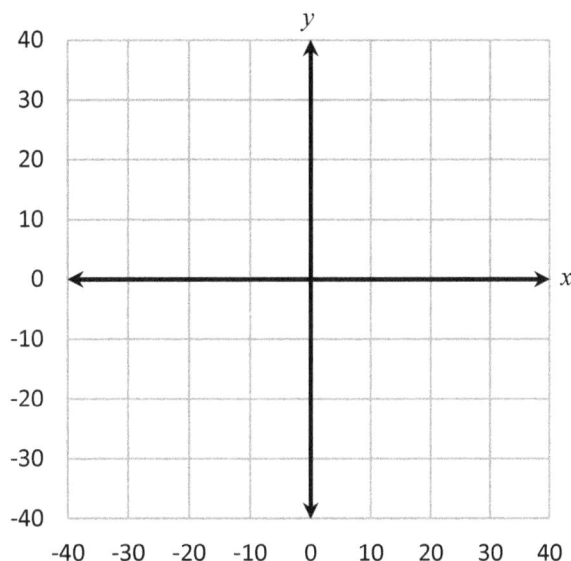

a. Jessica visits Bud's Bait and Tackle where Bud offers three types of rod/reel combinations. The table below lists the price and maximum casting radius (distance) in meters. Use the formula for the area of a circle $A = \pi r^2$ to find the available fishing area A, to the nearest meter, for each type. Then divide the price of each rod/reel combination by the fishing area A to obtain the fishing cost in dollars per square meter. Round your results to the nearest penny.

Rod/Reel Combo	Price (dollars)	Casting Radius r (meters)	Fishing Area A (square meters)	Fishing Cost (dollars/m^2)
Type 1	$15.95	10		
Type 2	$25.95	20		
Type 3	$49.95	40		

b. On the *xy*-plane above, depict the fishing area available for each rod/reel combination by graphing the three circles that represent the maximum casting radius for each rod/reel type. Use (0, 0) for the center of each circle. If you double the casting radius, do you get twice as much fishing area? Explain.

c. Write the equation of each circle that you graphed in part *b*. This represents the available fishing area based on the maximum casting radius for each rod/reel combination type.

   Type I: _____

   Type II: _____

   Type III: _____

d. Which type of rod/reel combination do you think is the best buy? Explain your reasoning.

e. Write the equation from part *c* that matches the "best buy" rod/reel type you chose in part *d*. Solve for *y* in terms of *x*. Is *y* a function of *x*? Explain your reasoning.

   Equation of "best buy" rod/reel combination: _____

f. What other factors should Jessica consider before buying a rod/reel combination?

1. Extension Exercise—Patterns

Sequences often come from observing patterns that occur in nature. Many chemical compounds exhibit natural patterns in molecular arrangements. In this activity, you will observe and extend some patterns, develop algebraic expressions for the general term of the sequence, and use the formulas you develop to predict what will happen if the patterns continue.

a. The first three cards of a pattern are shown below. Draw circles, squares, and triangles on the next two cards to continue the pattern. Then fill in the table below.

Card Number	Number of Squares	Number of Circles	Number of Triangles	Total Number of Shapes
1				
2				
3				
4				
5				
6				
7				
n				

b. Find an expression for the general term a_n of the sequence for the total number of shapes on the nth card. Use this general term to predict the number of shapes on the following cards:

$$= \underline{\hspace{6cm}}$$

Number of shapes on:

33rd card _____   75th card _____   150th card _____

c. Repeat part *a* for the following pattern:

Card Number	Number of Squares	Number of Circles	Number of Triangles	Total Number of Shapes
1				
2				
3				
4				
5				
6				
7				
n				

d. Find an expression for the general term a_n of the sequence for the total number of shapes on the nth card. Use this general term to predict the number of shapes on the following cards:

$$a_n = \underline{\hspace{5cm}}$$

Number of shapes on:

33rd card _____   75th card _____   150th card _____

e. Consider the following pattern of dots. Extend the pattern for cards 4 and 5. Then complete the table.

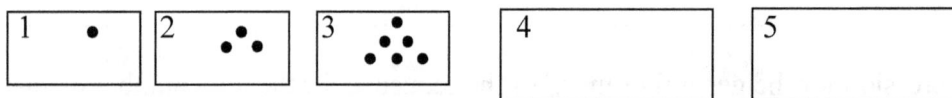

Card Number	1	2	3	4	5	6	7	n
Number of Dots								

f. Find an expression for the general term a_n of the sequence for the total number of dots on the nth card. Use this general term to predict the number of dots on the following cards:

$$a_n = \underline{\hspace{5cm}}$$

Number of dots on:

33rd card _____   75th card _____   150th card _____

2. Exploration Activity—Fibonacci and the Golden Ratio

The Fibonacci sequence of numbers starts with $a_1 = 1$ and $a_2 = 1$. Every other term of the Fibonacci sequence is sum of the two previous terms.

a. Using the definition above, find the next four terms of the Fibonacci sequence. Show the numbers added to get each term.

$a_3 = a_1 + a_2 = $ _____ + _____ = _____

$a_4 = a_2 + a_3 = $ _____ + _____ = _____

$a_5 = a_3 + a_4 = $ _____ + _____ = _____

$a_6 = a_4 + a_5 = $ _____ + _____ = _____

b. If two consecutive terms of the Fibonacci sequence are 610 and 987, what is the next number in the sequence? _____

c. Generate the first 50 Fibonacci numbers and write them in the table below.

What are a_{49} and a_{50}? $a_{49} = $ _____ $a_{50} = $ _____

n	a_n		n	a_n		n	a_n
1			18			35	
2			19			36	
3			20			37	
4			21			38	
5			22			39	
6			23			40	
7			24			41	
8			25			42	
9			26			43	
10			27			44	
11			28			45	
12			29			46	
13			30			47	
14			31			48	
15			32			49	
16			33			50	
17			34				

d. The number of ancestors of a drone bee forms a sequence with a distinct pattern. A drone is a male and has only one parent, the queen bee (a female). Each female bee (including the queen) has both a male and female parent. Complete the "family tree" of a drone bee up to six generations to verify that the number of ancestors at each level forms the Fibonacci sequence.

Total
Ancestors

6th level: ____

5th level: ____

4th level: ____

3rd level: ____

2nd level: (grandparents)   M   F   2

1st level: (parent is female)   F   1

Drone Bee is a male (M)   M   1

e. The Fibonacci sequence appears other places in nature and has a connection to the golden ratio, which is the number $\dfrac{1+\sqrt{5}}{2}$.

Find a decimal approximation for the golden ratio rounded to 6 decimal places.

$$\dfrac{1+\sqrt{5}}{2} \approx \underline{\hspace{4cm}}$$

f. In the Fibonacci sequence, if you divide any term by the one immediately preceding it (for instance, the third term divided by the second term), then the ratios get closer and closer to the golden ratio as you go further out in the sequence. Verify that the following ratios get closer and closer to the decimal approximation for the golden ratio above. For each ratio shown below, complete the ratio with the appropriate terms of the sequence and then give the decimal approximation rounded to six decimal places.

$\dfrac{a_4}{a_3} = \underline{\hspace{1cm}} \approx \underline{\hspace{1.5cm}}$ $\dfrac{a_8}{a_7} = \underline{\hspace{1cm}} \approx \underline{\hspace{1.5cm}}$

$\dfrac{a_{21}}{a_{20}} = \underline{\hspace{2cm}} \approx \underline{\hspace{2.5cm}}$ $\dfrac{a_{49}}{a_{48}} = \underline{\hspace{2cm}} \approx \underline{\hspace{2cm}}$

3. Conceptual Exercise—College Costs

According to The College Board, the annual costs for tuition, fees, room, and board at public and private universities alike have been increasing recently at an average rate of 2% per year. During the 2018-2019 academic year, the average annual cost of tuition, fees, room, and board was $21,370 at a public four-year university and $48,510 at a private four-year university.

a. In the table below, show what college costs would be if the rate of increase continues to be 2%. *Hint:* To compute a 2% increase, add the college costs for the year to 2% of the cost to get the next year's cost of tuition, fees, room, and board.

School Year	Public 4-year University	Private 4-year University
2018-2019	$21,370	$48,510
2019-2020		
2020-2021		
2021-2022		
2022-2023		
2023-2024		

b. Find the general term of the sequence that describes the pattern of average annual cost for public universities. Let $n = 1$ represent the 2018-2019 academic school year.

$a_n =$ _____

c. Find the general term of the sequence that describes the pattern of average annual cost for private universities. Let $n = 1$ represent the 2018-2019 academic school year.

$a_n =$ _____

d. Assuming that the rate of college cost increase remains the same, find the average annual cost for the 2033-2034 academic year for each type of university.

Public University: _____   Private University: _____

e. Use partial sums to find the average total cost of a four-year college education at a public university for a student starting in the 2023-2024 academic year. Show how you set up your expressions for calculations. (Use a calculator for the actual calculations.)

f. Use partial sums to find the average total cost of a four-year college education at a private university for a student starting in the 2023-2024 academic year. Show how you set up your expressions for calculations. (Use a calculator for the actual calculations.)

g. Explain why the above costs are only estimates. What other factors might affect the future costs of a student's tuition?

4. Group Activity—Investigating a Sequence

Materials needed: 2 index cards and a pair of scissors

a. Decide how to cut one index card so that it can be evenly distributed among three members of your group. What fraction of the card does each group member receive?

b. The remaining index card will also be distributed evenly among three group members. This time, however, the only allowable cut of the card is shown. Distribute three of the pieces equally among group members. Cut the piece left to make another distribution. Use the table below to record the fraction of the index card distributed to each group member at each step. Make at least four distributions, and then generalize the pattern to predict the remaining distributions.

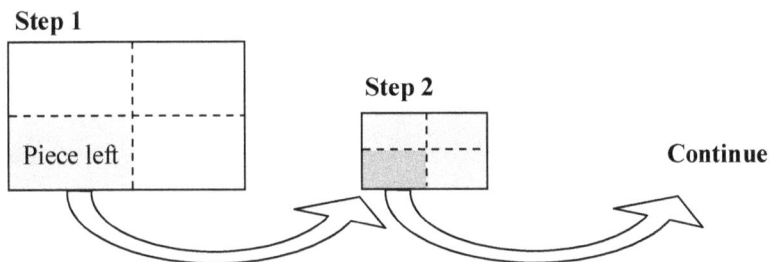

Step 1

Piece left

Step 2

Continue

Step	Fraction of Original Card Distributed	Fraction of Original Card Left	Step	Fraction of Original Card Distributed	Fraction of Original Card Left
1	$\dfrac{3}{4}$	$\dfrac{1}{4}$	4		
2	$\dfrac{3}{4} \cdot \dfrac{1}{4}$	$\dfrac{1}{4} \cdot \dfrac{1}{4}$	5		
3			6		

c. Write the first six terms of the sequence shown in the "Fraction of Original Card Distributed" columns in the table above, and write the *nth* term.

$a_1 = $ _____

$a_2 = $ _____

$a_3 = $ _____

$a_4 = $ _____

$a_5 = $ _____

$a_6 = $ _____

$a_n = $ _____

What type of sequence does this represent? _____

d. Based on the sequence you identified in part *c,* determine the value of the first term a_1 and the value of the common ratio *r*. Check your choices by showing that $a_2 = a_1 r^{2-1}$, $a_3 = a_1 r^{3-1}$, and $a_4 = a_1 r^{4-1}$ give the same results for a_2, a_3, and a_4 as found in part *c*.

$a_1 = $ _____ $r = $ _____

e. Use the formula $S_n = \dfrac{a_1(1-r^n)}{1-r}$ to find the sum of the first six terms of your geometric sequence. Show your substitutions. Simplify your answer to exact fraction form, then use your calculator to report your answer rounded to six decimal places. Compare your answer to the sum of the first six terms in part *c*.

$S_6 = $ $=$

_____   _____
     simplified fraction   decimal form

Sum of first 6 terms in part *c*:

f. Use the formula from part *e* to calculate S_8 and S_{10}. Round each sum to six decimal places. If you were to continue to calculate these partial sums S_n as the value of *n* gets larger and larger, what number do you think S_n will approach?

$=$ _____ $=$ _____

S_n will approach _____ as *n* gets larger and larger.

Verify your prediction by using the formula $S_\infty = \dfrac{a_1}{1-r}$ to compute the sum of the infinite sequence. $S_\infty = $ _____

Think about how your geometric sequence was formed by keeping track of the fraction of the original index card distributed at each step. Does the value of S_∞ that you just computed make sense in the context of dividing up the index card? Explain.